D0641816

Temalpakh

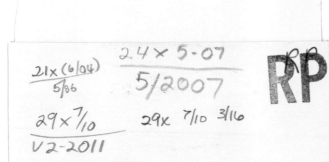

TEMALPAKH

(from the Earth)

Cahuilla Indian knowledge and usage of plants

Lowell John Bean and Katherine Siva Saubel

With an Appendix on the Problem of Aboriginal Agriculture
Among the Cahuilla by Harry W. Lawton and Lowell John Bean

Photographic Arrangement by Nancy Bercovitz

1972 MALKI MUSEUM PRESS, MORONGO INDIAN RESERVATION

Library of Congress Catalogue Card Number 72-85815
ISBN 0-939-046-24-5

First Reprint, February, 1979
Second Reprint, July, 1982
Third Reprint, June, 1987

Printed in the United States of America
Rubidoux Printing Company, Riverside, California

*Dedicated to the memory
of our mentors Juan Siva,
Calistro Tortes, Victoria Wierick,
Cinciona Lubo, and
Salvador Lopez.*

Contents

Illustrations

Illustrations (Continued)

Introduction

Southern California's native American Indian populations varied in historical background and in their social adjustments to the multifaceted environments in which they lived. On the coast, large sedentary populations made use of the combined resources of the sea, the coastal fringes, and adjacent inland plains and mountains. In the interior, various groups of the southern California basin skillfully adapted to a variety of ecological niches, developing a diversified hunting and gathering economy with far-ranging reciprocal trade relationships with other cultures. Far to the east along the Colorado River and its fertile shores, Indian cultures arose along more dramatically complex lines than in much of aboriginal California. Almost midway between the extremes of the rich coast and interior basin and the hospitable Colorado River area lay the territory of the Cahuilla Indians. They inhabited an environment beginning west of the mountainous terrain that formed an eastern wall to the interior basin and extending far out into the Colorado Desert, seemingly harsh and barren, yet richer in natural resources than the unknowledgeable might imagine. Here the Cahuilla Indians—whose oral literature reflects a land of sharp contrasts— became singularly adept at mastering and exploiting an environment that was both desert and mountainous woodland.

The Cahuilla are a Shoshonean-speaking people whose historical relationships connect them with the well-known Hopi of Arizona, the powerful Gabrielino on the Pacific Coast, the Luiseño who extended into the interior basin, the Diegueño to the south, and many desert-oriented groups of southern California, including the Serrano, Kamia, Chemehuevi, and Paiute. Before contact with the Spanish, the Cahuilla population may have been as high as 6,000 persons, spread over 2,400 square miles and divided into seven or more politically autonomous groups. Each group was composed of from four to approximately a dozen village land-holding units (lineages) with villages scattered throughout the area bounded by the San Bernardino Mountains on the north, the middle of the Colorado Desert to the east, the Anza-Borrego Desert and Santa Rosa Mountains to the South, and the San Jacinto Mountains and portions of the San Jacinto and San Bernardino valleys to the west.

Long before arrival of the Spanish in California, the Cahuilla had acquired mastery of a seemingly forbidding and fruitless environment. As with most hunting and gathering peoples, their daily life was influenced to a great extent by the environmental potential which nature afforded them. No one can attempt to understand the Cahuilla without soon becoming aware that they were master ecologists—that in their world view and philosophical system there was an ecological model as keenly geared to empirical reality as any that botanists might teach in today's college classrooms.

The historical and cultural backgrounds of the Cahuilla have been written about by early explorers and by historians, anthropologists, and fascinated laymen for more than a century, beginning with the earliest description of these people by Juan Bautista de Anza on his first trek through their country from Sonora to San Gabriel in 1774. The literature of California abounds in references to the Cahuilla, and a bibliography published by Malki Museum Press (Bean and Lawton 1967) cites more than 500 pertinent sources of material.

The first major publication on the Cahuilla has long been a classic in American anthropology: David Prescott Barrows' *The Ethno-botany of the Coahuilla Indians of Southern California* (Chicago, 1900). Barrows' work, which laid the foundation for all future Cahuilla studies, was primarily concerned with plant uses, although it also touched upon the history, religion, language, and other aspects of Cahuilla culture.

Temalpakh, which is the Cahuilla way of saying "from the earth," draws upon Barrows' pioneer research, other more recent sources, and more than ten years of ethnobotanical field work by the authors. Our purpose remains essentially the same as Barrows' original intent—to understand how the Cahuilla adapted to their natural environment. This book approaches the subject in greater depth, however, and is oriented strictly to plants, their multiple uses, and their influence on the culture of the Cahuilla. Barrows gathered information on the use of about 100 plants. This book lists more than twice that number and examines plant species in fuller detail with consideration of their various uses, distribution, dependability as food sources, methods of gathering, and preparation and storage procedures.

The basic pattern of our study is explained in the "Ethnobotanical Report Sheet" (Bean 1961), which was used both in our literature survey and in field interviews as a guide in our preliminary work. Briefly, the report sheet was concerned with what plants were gathered, how they were gathered and prepared, and in what ways they were used. In addition, available data was obtained on the nutritional value of plants, their dependability or seasonal availability as a food source, and whether they could be preserved by storage. Other areas of interest were division of labor and expenditure of time in gathering, effects of plants on settlement patterns, plant associations with religious ceremonies, and trade with other Indian groups.

Data gathering for this work began in 1960 when Lowell Bean, the senior author, was residing with Mr. and Mrs. Elmer Penn on the Morongo Reservation near Banning, California. At that time, he and his co-author, Mrs. Saubel, who has had a lifelong interest in the ethnobotany of her people, commenced collecting data, which they continued to obtain intermittently over the years. The bulk of this material was acquired directly from Cahuilla people, although published sources were also consulted and are cited wherever information from them is used in the text.

Our field technique was to collect plant specimens and offer them to individual Cahuilla for examination and comment. Sometimes plant samples were taken to Cahuilla homes for inspection, but whenever possible Cahuilla people were taken into the field so they could select plant samples for themselves. The latter approach was most productive, since it not only saved considerable time but also often served to refresh the memory of those assisting us concerning gathering techniques or uses of a plant. Once plants were clearly identified, all information was checked whenever feasible with at least several other Cahuilla people so that individual variations in their observations could be noted and errors corrected.

Some peculiarities of our text deserve to be explained. A number of Cahuilla plants are listed which were presented in an ethno-botany by Romero (1954), but which were not recognized by any of the Cahuilla with whom we worked. In many cases, there were plants for which no Cahuilla name could be obtained, usually because our sources had forgotten the Indian nomenclature after so many years of speaking Spanish or English almost exclusively. About one hundred plants were remembered as having been used by the Cahuilla which we were unable to identify botanically. In the case of plants that were mentioned which we could not locate in the field, we decided to publish the meager information available in the hopes that someone may eventually be able to identify these plants. There are also a number of plants discussed which were remembered by only one person. In such instances, we have recorded the name of the person who provided the information.

Obviously, our book is not definitive, since despite the many years that have passed since it was begun and the many people who have worked with us, it can represent only a shadow of the original knowledge held by Cahuilla scholars of the past. Yet when one considers how much material it has been possible to obtain from a few Indian people so distant in time from their ancestors, there can be few remaining doubts about the highly sophisticated knowledge of the environment possessed by the Cahuilla.

The authors hope that this work will stimulate a better understanding of environmental and cultural relationships which will be at least partly applicable to all other southern California Indian groups. Although the Cahuilla are the only Indian group in southern California who remember so clearly their original subsistence patterns, their plant domain is sufficiently similar to that of neighboring tribes that valuable comparisons can be made. There still exists an enormous body of unpublished information on plants and their uses among such neighboring groups. Two researchers, the senior author and Charles Smith, currently are compiling material for a work that will describe over 500 plants and their uses among twelve southern California tribes extending as far north as the Owens Valley Paiute and as far south as the Mexican border.

This book and such future ethnobotanical work should provide anthropologists and archaeologists interested in exploring the dynamic relationship between culture and environment with a rich body of data which can be used as new theories and hypotheses are developed that take account of botanical variables. In addition, we hope that this work provides the lay reader with a keener appreciation of the ecological dependence of man on his environment and a heightened awareness of the detailed, precise scientific knowledge that the

California Indians had of the world around them. The reader who wishes to know more about the Cahuilla, especially such information as an explanation of the status terms used in this book, is referred to Lowell Bean's recently published *Mukat's People: The Cahuilla Indians of Southern California* (Berkeley, 1972).

The authors have dedicated this book to five Cahuilla friends who provided much information for the work and who had looked forward to its completion: the late Juan Siva, Calistro Tortes, Salvador Lopez, Victoria Wierick, and Cinciona Lubo. Others who contributed from their immense fund of knowledge or assisted us from the beginning of our research were Mrs. Alice Lopez; Mrs. Jane Penn, curator of Malki Museum, Inc.; and Mariano Saubel. We also are very grateful for contributions made by William Holmes, Matthew Pablo, Robert and Esther Levi, Tom Siva, Alex Siva, and Alvino Siva.

Many others close to the Cahuilla helped support our efforts. Initially, the research was inspired by suggestions and the sponsorship of Dr. Wendell Oswalt of the University of California, Los Angeles. Mr. and Mrs. Harry C. James shared their knowledge of Cahuilla culture and the hospitality of their lodge at Lake Fulmore. Mr. and Mrs. Francis Johnston accompanied the authors on many field trips, and we drew constantly upon their intimate knowledge of the desert and Cahuilla Indian sites. Mrs. Edna Badger deserves special mention for help in locating various plants and other efforts on our behalf. Dr. C. E. Smith, who was director of Palm Springs Desert Museum when the senior author was a curator at that instiution, placed museum facilities at our disposal, and was a steady source of encouragement. We are grateful to both Dr. Smith and the Board of Directors of Palm Springs Desert Museum.

Among those to whom we are obligated for professional advice are Dr. William Bright, Dr. Richard Logan, and Dr. Mildred Mathias, all of the University of California, Los Angeles; Dr. Hansjakob Seiler of the University of Cologne, Germany; Dr. Margaret Langdon of the University of California, San Diego; Dr. Philip Munz and the staff of the Santa Ana Botanical Gardens; Florence Shipek of San Diego; and Beecher Crampton, lecturer in agronomy, University of California, Davis. Miss Christine Hays assisted in making plant identifications and collating data. Finally, we wish to express our special indebtedness to Oscar Clarke, herbarium botanist of the University of California, Riverside, who devoted many long hours to helping identify plants, discussing problems we encountered, and offering suggestions that have immeasurably strengthened this book.

Mrs. Nancy Bercovitz took many of the photographs in this volume and was responsible for photographic layout. Her photographs in this work represent only a fraction of many valuable pictorial studies she is making of Cahuilla culture. The Southwest Museum of Los Angeles, the Lowie Museum of the University of California, Berkeley, and Palm Springs Desert Museum have generously assented to use of historical photographs from their files. We are deeply grateful to the American Indian Historical Society for permission to reprint the article on aboriginal agriculture by Harry Lawton and Lowell Bean, which appears as an appendix to this volume, and which was first published in *The Indian Historian*. Editorial suggestions at various stages of manuscript preparation were made by Dr. Hansjakob Seiler, Edna Badger, Jean Marie Taylor, Charles Smith, and Robert Gill. The final manuscript was reviewed by members of the Malki Museum Press editorial board and Oscar Clarke. We are particularly indebted to Harry Lawton for editing the work and guiding it through publication.

Riverside Municipal Museum, its board of directors, museum association, and staff under the direction of Charles Hice made a significant contribution to completion of this book. Originally, the museum offered the authors an opportunity to publish the work and provided some research funds. This aid came at a critical period and is much appreciated. We are also grateful to the Faculty Research Committee, California State University, Hayward, for a grant to carry out field research. Finally, we wish to thank Wenner-Gren Foundation for Anthropological Research and the American Philosophical Society for providing part of the funding for studies of southern California Indians.

The authors recognize that there is currently much interest in native American food plants, especially among young people fascinated with the idea of living in an ecological balance with nature. Some of the plants discussed in this book occur only sparingly in the environment today, and should not be disturbed since widespread collection could result in their extermination. We have tried to alert readers to all plants known to be toxic, but it must be emphasized that all unfamiliar plants should be treated with cautious respect. Readers should remember that the Cahuilla had centuries for experimentation with plants in which to become familiar with their risks and rewards. Some toxic plants were used for food and medicine by the Cahuilla, but they were well acquainted with the proper amounts to use or processing methods which reduced toxicity. The information gathered here is often fragmentary, representing only a fraction of original Cahuilla plant knowledge. For that reason, readers should recognize that much more knowledge of the Cahuilla past would be needed to safely use many of the plants discussed.

—Lowell John Bean
Katherine Siva Saubel

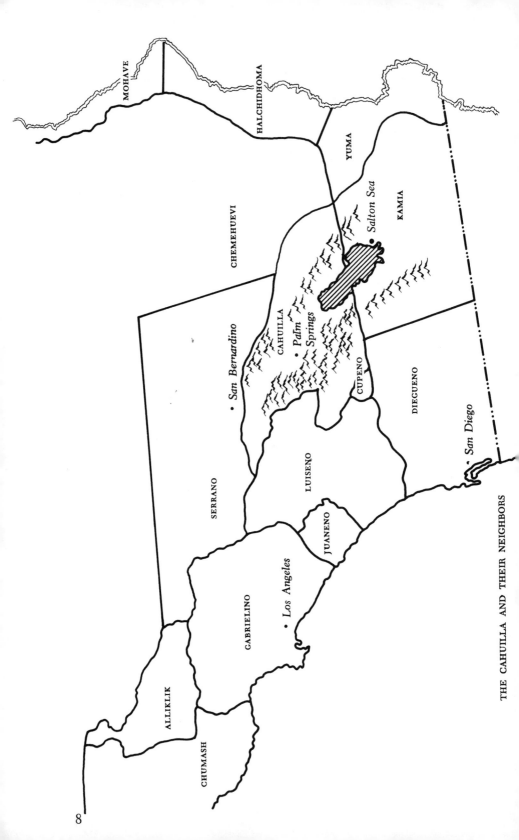

THE CAHUILLA AND THEIR NEIGHBORS

MOHAVE

HALCHIDHOMA

YUMA

CHEMEHUEVI

KAMIA

Salton Sea

CAHUILLA

• San Bernardino

• Palm Springs

CUPENO

DIEGUENO

SERRANO

LUISENO

• San Diego

JUANENO

GABRIELINO

• Los Angeles

ALLIKLIK

CHUMASH

8

The Cahuilla Natural Environment

The territory occupied by the Cahuilla is unusual geographically in its great variety of topography, climatic conditions, and biological life zones. These circumstances directly influenced the nature of the plant community upon which the Cahuilla depended for subsistence, and the plant community, in turn, significantly affected development of Cahuilla settlement patterns, social structure, ritual, and philosophy (Bean 1972).

The northern boundary of Cahuilla land is marked by the southern slopes of the San Bernardino range and its smaller extensions to the east, the Little San Bernardinos and the Cottonwood and Eagle ranges, which separate the Mojave Desert from the Colorado Desert. These mountains served as a recognized boundary between the Cahuilla and their northern neighbors, the Serrano Indians. The southern foothills of these mountains and the intervening San Gorgonio Pass, which separates them from the San Jacinto Mountains, were extensively occupied and exploited by the Cahuilla. Towering above the San Gorgonio Pass is Mount San Gorgonio (11,502 feet), the highest peak of the San Bernardino range, and the highest mountain south of the Sierra Nevadas.

South of the Little San Bernardinos and east of the San Gorgonio Pass lies the Colorado Desert. This includes an ancient lake basin, one of the lowest points in the United States, ranging in some areas to 300 feet below sea level. The western edge of this desert is shadowed by the San Jacinto and Santa Rosa Mountains, which receive so much precipitation that they support fine forests and can in no way be considered desert areas. The higher rainfall near these ranges was reflected in a more intensive use and occupancy by native Californians than in the more easterly and arid sections of the Colorado Desert. From the steep slopes of the San Jacintos, surmounted by Mount San Jacinto (10,786 feet), the desert floor descends sharply in a few brief miles to 300 feet below sea level— one of the swiftest topographical ascents in North America.

The San Jacintos, which lie at a right angle and to the south of the San Bernardinos, provided the principal food-gathering ranges for the Cahuilla. This is an area of relatively high rainfall conducive to rich vegetation. Contiguous to the San Jacintos and running easterly are the somewhat drier Santa Rosa Mountains, which descend abruptly to the floor of the Anza-Borrego Desert on the southeast. This arrangement of desert adjacent to mountains provided the Cahuilla with a contrasting and advantageous variety of life zones. The Cahuilla on the southern slopes, residing in Los Coyotes and Rockhouse canyons, customarily exploited both the desert and mountain terrains as did Indians on the western edge of the Colorado Desert at Palm Springs and Indio. West of the San Jacintos, there are rolling foothills and valleys with plains typical of the San Jacinto Valley. The Cahuilla inhabitants of this area benefited from higher rainfall and a richer, although less varied vegetation than the desert-oriented Cahuilla.

Average daily temperatures within the Cahuilla territory vary from 40°F to 52°F in January and from 75°F to 100°F in July, although there are extreme seasonal variations and temperatures can become very hot in the desert areas during the summer, sometimes reaching as high as 125°F. Judicious selection of village sites took into consideration the possible ranges of temperature so that people were as comfortable as possible throughout the year. Thus villages were usually placed on alluvial fans and at canyon mouths where they received the benefit of warm winter breezes from the desert and cool summer winds coming down the canyons from the mountains.

Rainfall varies radically from one area to another in Cahuilla territory. It ranges from 4 inches or less annually on the desert to 30 inches or more in the mountainous areas. The problem of water supply was solved in drier areas by the hand-dug water wells for which the desert Cahuilla were renowned. In addition, permanent springs and streams which emerged out of the mountain ranges onto the foothills and the alluvial fans helped provide the Cahuilla with sufficient water. Although many of the Cahuilla were desert-oriented, water does not appear to have been a particularly serious problem, except during extremely dry years.

BOTANICAL LIFE ZONES

Using the life zone classification of Hall and Grinnell (1919: 37-44), we shall describe the major types of botanical life forms available to the Cahuilla. Even in these zones there is considerable variability in patterns of plant distribution, and life zone models should be viewed as an ideal rather than a real depiction of flora zonation. The actual distribution of plants, although usually fairly consistent with the life-zone model, varies frequently enough so that factors causing such variability should be pointed out. Often plants which are typical of one zone may be found in another because of one or more of the following factors: (1) drainage of cold air from hillslopes, producing warmer thermal belts there; (2) accumulation of such cold air in valley bottoms, producing areas of exceptional coldness; (3) streams carrying cold water that cause higher altitude plants to grow at lower elevations than might be generally expected; (4) evaporation from moist soils, and lingering snow banks that depress temperature in some areas and change the botanical pattern; (5) rocky slopes and outcroppings that tend to be warmer than surrounding areas and thereby cause differences in plant formation; and (6) contrasts in exposure to solar radiation, producing warmer but drier south-facing slopes and cooler but more moisture-retaining north-facing slopes. These factors all work to frustrate any attempt to construct a neat, consistent pattern of Cahuilla vegetation resources. To demonstrate the range of resource variability however, we will present a brief summary of life zones borrowed from Hall and Grinnell's typology as a useful guide for the reader in keeping in mind the types of environment in which various plants grow.

Life zones in Cahuilla territory can be classified into four categories: Lower Sonoran, Upper Sonoran, Transitional, and Canadian-Hudsonian. Each of these zones provided a distinctive Cahuilla food collection potential.

LOWER SONORAN LIFE ZONE

The Lower Sonoran Life Zone, which ranges from the desert floor to 3,500 feet, lies generally below the pinyon-juniper belt.

This zone is characterized by low rainfall, averaging about 4 inches annually on the desert floor, fine-textured alluvial to sandy or gravelly soils, and xerophytic plant communities. Over the broad desert flats, the creosote bush is the dominant plant, replaced by the saltbush in areas of more saline or alkaline soils. Washes are characterized by desert willow, smoke tree, palo verde, ironwood, and catclaw, and areas with water at or near the surface by California fan palm, mesquite, screwbean, arrowweed, and reeds. Well air-drained slopes, particularly with southern exposures, have higher temperatures on critically cold winter nights and support frost-sensitive plants such as cholla, barrel cactus, ocotillo, century plant, and creosote bush. Higher elevations and north-facing slopes, both with cooler conditions, are the locale of the Mojave yucca, goatnut, and nolina. Thus, even within the Lower Sonoran zone there are distinctive plant communities.

Approximately 60 percent of the land within Cahuilla territory lies inside the Lower Sonoran zone. The authors have tentatively estimated that more than 40 percent of the plant species exploited by the Cahuilla may be found in this biotic region, particularly on alluvial fans. This estimate includes those plants characteristic of other life zones which overlap into the Lower Sonoran zone. The food resource value of plants in this zone was particularly high, because as Aschmann (1959b:43) noted: "Desert plants in general spend far more of their energy on reproduction and so provide far more concentrated nutrients suitable for animal consumption than would the same amount of vegetative matter grown in more favored regions." This is particularly true of annuals, which concentrate on seed reproduction for survival. In contrast, perennials on the desert may produce no seed in low rainfall years.

UPPER SONORAN LIFE ZONE

The Upper Sonoran Life Zone extends upward to the 5,000-foot level. Summers are usually warm in this zone and winters are cold with rainfall averaging about 15 inches annually. On the desert slopes, pinyon and juniper are the most characteristic species, although there are notable exceptions such as a stand of ribbonwood or red shank at the southern end of Pinyon Flats and a belt of chaparral on the northern slopes of the Santa Rosas. On the western slopes of the San Jacintos and in the foothills of the San Gorgonio Pass, however, chaparral belts are dominant. Although pinyon and juniper are usually thought of as forming a plant association, juniper actually occurs at a slightly lower elevation and interlaps with pinyon. Other characteristic plants of the Upper Sonoran zone are chamise, ironwood, oak, ocotillo, manzanita, buckthorn, and barrel cactus.

More than 30 percent of the land available to the Cahuilla lay within this life zone. The authors estimate that over 60 percent of the food plant species used by the Cahuilla—including those species which overlap from other zones—may be found in this life zone.

TRANSITIONAL LIFE ZONE

The Transitional Life Zone ranges from 5,000 to 7,000 feet and sometimes slightly higher. This biotic zone is characterized by relatively cool summers and cold winters with an annual precipitation of 20 to 30 inches. These were the useful forested areas of Cahuilla territory. The forests are coniferous and the yellow pine group is dominant. There are scattered oak groves throughout this life zone, and along streams cottonwood and willow trees are common. In other areas, a chaparral type of vegetation may occur. Abrams (1910:310-312) recorded some fifty plant species typical of this zone. For the Cahuilla, the most important plants were oak (particularly *Quercus kelloggii*), manzanita, and elderberry.

The authors estimate that about 15 percent of the plants utilized by the Cahuilla, including overlapping species more characteristic of other zones, lay within this region. The importance of this zone is magnified, however, by the presence of California black oak, one of the most favored of Cahuilla staples.

CANADIAN-HUDSONIAN LIFE ZONE

In this zone, which extended from 7,000 feet to the summits of the mountain ranges, there was very little plant life to be exploited by the Cahuilla. Rainfall is heavy in this zone, and during the winters subfreezing temperatures may occur for long periods. During the summer, this area was primarily exploited by hunters in search of large game. Occasionally, the few useful plants that grew at these high elevations were gathered. Among the characteristic trees of this zone is the lodgepole or tamarack pine. This biotic zone comprised about five percent of Cahuilla territory.

Readers interested in a more thorough discussion of the environmental factors affecting Cahuilla subsistence and culture will find such information in *Mukat's People* (Bean 1972).

Plants And Cahuilla Culture

Among the Cahuilla plants were not viewed simply as objects which might or might not be useful to man, but as living beings with whom one could communicate and interact. Plants were one of a number of life forms such as rocks, elemental forces, animals, birds, and spirits that could communicate with those who knew how to "listen." Plants play their own anthropomorphic roles in Cahuilla traditional lore and in the epic-length cosmogony and other oral literature. This role of plants was not simply metaphorical or literary, however, since Cahuilla oral literature was considered to be an accurate representation of past events and of a natural reality that was on-going in the present.

Plants were placed on earth by the Creator to serve man, but it was not intended that this should be an exclusive one-way relationship. Man was seen as one of a number of cooperating beings who shared in the workings of a universe that was an interacting system. All parts of this system were reciprocal and man had obligations to the rest of the universe and its creatures. Plants, like any life form, were therefore treated with respect. A person gathering a plant would thank the plant for its use, apologizing in one sense for the harm inflicted on the plant, but also recognizing that it was natural that the plant submit to its predetermined use. In the Cahuilla cosmogony or creation myth, the god *Mukat* decrees that the first fire will be made out of *Ninmaiwit*, the woman who is a palm tree. Her screams of anguish are pitiful, yet the people recognize the inexorability of her sacrifice as part of the order of the universe.

Each use of a plant, however, required an acknowledgement of the indebtedness of the user, and this was expressed in rituals associated with such use. These rituals were a recognition that there was a right conduct associated with the use of any life form, and they also helped to ensure regeneration of the life form. One of the fundamental obligations of each user was not to overexploit a plant, so that its survival was endangered. For this reason, when plants were collected, a part of the plant was usually left behind and it was seldom stripped bare. Nor were all plants of one type in any gathering area ever collected; some were always left behind for species propagation.

There were not only rituals accorded to the plant, but there were also rituals that extended beyond the plant as an individual life form to those supernatural agencies responsible for plant fertility. A first-fruit ceremony was therefore held at the beginning of the gathering season for most major foods such as agave, mesquite, acorn, and pinyon. Lineage members of a village gathered at this ceremony to eat a ritual portion of the crop. The ceremony lasted three days and three nights and was held to express appreciation to the supernatural powers for providing a bountiful harvest. The ceremony also ensured that any "sickness" which was put into the food at the time of Creation was exorcised and that the supernatural beings responsible for plant fertility would provide crops in future seasons.

CAHUILLA FOLK TAXONOMY

Cahuilla folk taxonomy was extremely sophisticated, and the names given to plants were very exactly chosen and are often in keeping with modern botanical classifications. Each species of oak and yucca recognized by botanists, for example, is also classified by the Cahuilla. Occasionally, plant families were also recognized: all cacti or members of the family of Cactaceae are encompassed by the Cahuilla term *navtem*. More general categories were also recognized: a number of genera of berry-producing plants, for example, were lumped together under the term *piklyam*. The finiteness of categorization was usually related directly to the importance and extent of plant usage.

Generally, the Cahuilla identified plants by outward appearance, using such criteria as the morphology of plant parts. Whether or not plants had a human use was also important. Non-useful plants were often given a purely descriptive name, such as "hummingbird's tongue" or "wind-pet." Names of useful plants frequently reflected their function. Many plants fell into descriptive categories such as "fuzzy," "thorny," and "waxy" or "non-waxy." Smell was an additional significant criterion in plant classification. Almost every elderly Cahuilla when asked to name a plant regards smell as a primary observation to make before giving an identification.

Cahuilla plant knowledge—even in the fragmentary lore that survives—is so broad and intimate that one wonders how the Cahuilla discovered so many diverse uses for plants. There must have been thousands of experiments over centuries to arrive at so many plant uses. Some hint of this process may be found in a notebook left by the late William Pablo, a *puul* (shaman) who was still keeping empirical records of his plant findings in the early part of this century. New foods and medicinal plants were experimented with cautiously in sparing quantities before being introduced to general use. It was the *puvulam* (shamans) and older women, especially those who were *tingavish* (doctors), who concerned themselves with botanical knowledge and experimentation.

The Cahuilla began learning about introduced plants from the whites soon after contact. They were wary of new products, however, partly because they mistrusted white men and also because many European diseases accompanied white arrival. At the same time, a people so versed in ecology could not resist the challenge of new plants. The new plants were adopted into the Cahuilla culture, but only after careful experimentation. Certain new plants, particularly various cereal grains, appear to have been introduced swiftly, possibly because of their botanical similarity to many familiar native grasses. Other new plants, such as potatoes, tomatoes, and cabbages, appear to have been adopted by the Cahuilla relatively late in the nineteenth century, even though they were grown as staples at the Spanish missions along the coast. One rule-of-thumb in trying out new plants was to observe whether they were eaten by birds and animals. Another rule-of-thumb, recalled by Mrs. Alice Lopez, related to plant color. She remembered, for example, that white-colored berries were usually avoided.

The Cahuilla and Southern Diegueño (including the related Kamia) appear to be the only Indian groups in southern California who possessed native words for certain cultivated plants and a terminology suggesting the existence of concepts related to agricultural techniques. A comparative study of southern California Indian myths made by Harry Lawton (1967) indicated that crop plants figured only in the myths of these groups. The crop plants predominantly mentioned in Cahuilla myths were those which were part of the aboriginal agricultural complex of certain Colorado River tribes to the east of the Cahuilla. These facts and other circumstantial evidence studied over the past five years suggest that agriculture may have reached the Cahuilla by diffusion from the east prior to Spanish contact. The case for aboriginal agriculture among the Cahuilla is presented in the Appendix.

The pervasiveness of Cahuilla concern with plants is reflected in many ways in their culture. Names for some of the seasons referred to growth stages of plants. Plant names were often used in naming people, particularly girls. Cahuilla lineage names were occasionally suggestive of plant relationships. Many place names gave clues to dominant plants of a locale. The latter usage was frequently functional. When a Cahuilla knew the dominant plant of an area, he was skilled enough as an ecologist to immediately predict or anticipate what other plants and animals might be part of that area's ecosystem.

PLANTS AND SOCIAL ROLES

Cahuilla custom governing the division of labor provided that the gathering of food plants was generally an activity of women, but not their exclusive province. Since men through hunting often became familiar with locations of the best stands of plants, no actual restrictions were placed on what plants they might collect or process.

In the case of some food plants, such as agave, men were almost the exclusive collectors because of the physical labor involved in gathering. A few major crops such as pinyon, acorn, and mesquite involved the labor of the entire family. Through ritual, men also played a primary role in all plant gathering, since they were responsible for maintaining supernatural support of the ecosystem and indicating when plants were ready to be gathered and who should gather them. Although excluded from many formal rituals and political roles in aboriginal times, women always constituted a powerful informal political force, maintaining a valued position in their husbands' community as well as in their own community of birth.

Most Cahuilla were very familiar with the uses of the majority of plants, particularly those used for food. Certain people, however, because of their special training, knew more about specific plants than others. Among such people were shamans, doctors, and basket-makers, all of whom had to be intimately acquainted with specialized plant usages.

Cahuilla medical knowledge, more than any other area of plant use, was the province of specialists. *Puvulam* (shamans) and *tingavish* (doctors) made special studies of plant uses, experimented with various treatments, and learned the precise dosages of plant substance required to effect cures. Although some of this knowledge was public, much of it was kept guarded and passed on only to apprentices and friends. Shamans and doctors met frequently and compared curing techniques. At the same time, some specialists kept secret particular remedies.

The shamans occupied a more prominent position in the community than did the doctors. They employed both sacred power and plant remedies in curing as opposed to the doctors who relied on herbal treatments. The differences in status between these two roles have been discussed by Bean (1972). Doctors were usually women renowned for their curing abilities. This role along with basketmaking was one means by which women could acquire unusual stature in a male-dominated culture.

Equipment of shamans and doctors comprised special grinding equipment such as mortars, cutting implements (for surgery), and special pots for cooking medicinal plants. In addition, shamans also employed certain "power-enhanced" objects to cure patients. Mortars and pots of the shamans were frequently endowed with powers emanating from *nukatem* (spiritual guardian beings). Other equipment having power included special rocks and wooden wands to which feathers (usually owl) and sometimes an eagle claw or a raven's wing were attached. Elaborate song and ritual accompanied cures performed by shamans as well as the use of the power-endowed objects in various ways. Among medical treatments employed by shamans were herbal remedies, sweating, massage, prescribed dietary restrictions, and sucking and the laying on of hands.

THE SEASONAL ROUND

Each Cahuilla lineage had a permanent village judiciously located in relation to the natural resources of an area. There was no time when an entire village moved to another location, although considerable numbers of people within a village might move to other areas at certain times of the year in following the seasonal round of gathering activities. Villages were usually situated in the lower part of the Upper Sonoran Life Zone within the center of the richest food-gathering area. As food ripened in different areas, individuals and groups moved out from the village to harvest the crops. No village was located more than sixteen miles from all of its food-gathering ranges, and approximately 80 percent of all food resources used by a village could be found within five miles. As a consequence, no major population movement was necessary for subsistence.

In the early part of the year from January through February, there was little gathering activity except for agave, which had begun ripening in November. Groups of adult men and young boys left the village for several days to gather and prepare agave. Since there was little fresh food to gather, there was heavy reliance on food stored from the previous year. Game was plentiful, however, since animals had descended to lower elevations where the climate was more agreeable.

With the onset of spring, the desert and hills began exploding with numerous ripening plants. Many green fruits and buds became available within a short walk from the village, often within a few hundred yards and rarely farther than a mile or two. Yucca, wild onion, barrel cactus, tuna cactus, goosefoot, catclaw, and ocotillo were among the many plants harvested for their edible parts from April through May.

With the arrival of summer, botanical resources continued to expand through the months of June and July, and many plants characteristic of the late spring were still available. Staple tree fruits such as the honey mesquite and screwbean were ready for gathering in immense quantities. People living in the Upper Sonoran Life Zone areas such as the San Gorgonio Pass or the foothills bordering the Borrego Desert now moved down into the Lower Sonoran desert floor areas to harvest these crops. These were not movements of entire villages, but a large number of people did participate to ensure rapid gathering of the pods in a quantity that could be stored by the village for as long as a year. In addition, within three or four miles of each village there were abundant foods in the foothills, including manzanita and many other berries, yucca, and various *Opuntia* cactus species. The summer months were perhaps the busiest gathering period of the year, but collecting was mainly carried out by small family groups, usually women.

Through August and September another series of foods became available, and many foods of the late summer could still be gathered. Fall foods included numerous grass seeds, chia, saltbush seeds, pinyon nuts, palm tree fruit, thimbleberry, wild raspberry, wild blackberry, juniper berry, and chokecherry. Many of these food plants were at higher elevations, but they were still within a few hours of each village. Pinyon and juniper trees were often situated ten miles or more from villages, and family groups went to these groves to work as a team, returning home usually within a week with a maximum harvest.

The last important gathering season of the year was in October and early November when the acorn ripened. The acorn harvest was a great occasion for Cahuilla villages. The bulk of each village marched to their oak groves, generally within five or ten miles, and camped to gather acorns. Such collection trips lasted from one to three weeks, and acorns were hauled back to the village by the ton. During this period, a number of mountain berries also matured, including juniper mistletoe and California holly.

By late November gathering slowed down and there was time for a period of winter ceremonials. These religious observances which had been decreed by the creator-god Mukat served many functions, particularly the regeneration of the universe for the year ahead. By late December, the men were again seeking out agave plants and a new seasonal round had begun to replenish the village storehouse.

CAHUILLA PROPERTY AND OWNERSHIP CONCEPTS

Exploitation of plant resources was closely regulated by the Cahuilla. Not only were ritual restrictions imposed by Cahuilla leaders on who could gather plants and when they could be gathered, but regular "taxes" were levied against those with excessive reserves so that all community needs could be met as they arose. These taxes were regulated both by the ritual system and through regulations surrounding in-laws and marriage relationships (see Bean, 1972).

Another means of controlling the use of the environment was through resource ownership. Families were guaranteed perpetual gathering privileges and thereby economic security through individual ownership of oak trees, mesquite groves, or other plant stands. The extent and type of ownership related to the kind of plant, its ease of access, predictability of a crop, and the amount of food a plant produced.

Oak trees and mesquite which produced regularly and in great quantity were privately owned. Food sources which produced less predictably, such as pinyon, were generally available to various people on a first-come basis. Certain food producing areas, such as canyons with large stands of cacti, were owned by lineages with gathering in such areas regulated by Cahuilla leaders. Each lineage had land in three categories of ownership: (1) land that was privately owned by lineage members; (2) land owned by extended families; and (3) land owned by the lineage in common. In the latter areas, any lineage member could collect plants. Since lineages were also members of larger corporate groups (sibs), there was also land open and available to members of several lineages belonging to a sib.

Conflicts between Cahuilla most commonly occurred as a result of one group trespassing upon another group's property. Theoretically, conflict became necessary only during poor seasons when there were food shortages. Nevertheless, individuals did occasionally trespass on gathering areas of other groups, and punishment or demands for reparation had to be carried out immediately to preserve property rights.

A number of categories of land ownership related to specialties. A shaman might own an area where there was a stand of medicinal plants. Basketmakers owned special areas where plants grew that provided materials for basketmaking. One unusual type of land ownership was the individual ownership of woodpecker-infested trees. Woodpeckers customarily store nuts of various kinds in holes in trees. Such trees could be claimed by the first man who found them and incorporated into his own private collecting area. All privately owned gathering areas were referred to as *kiiwut* or "place you are waiting for."

TRADING OF PLANT PRODUCTS

Since distribution of plants is directly related to microenvironmental differences in Cahuilla territory, plants were a frequent item of exchange between families in a village, and between families across village, sib, and tribal lines. Cahuilla people customarily traded plant food products with one another and with tribes as far west as Los Angeles, southward to the San Diego area, and as far east as the Colorado River.

Trading between distant tribes was for plant foods or other

subsistence resources that were scarce or unavailable in Cahuilla territory. Often trading was conducted for treasured goods such as ritual equipment, jewelry, baskets, obsidian, and other manufactured articles. When a group had a surplus of any perishable product, it invariably sought to exchange it with other groups for non-perishable items. Trade occurred throughout Cahuilla territory and between surrounding tribes on a grand scale within a ritual exchange system that has been described in greater detail in *Mukat's People* (Bean, 1972).

CAHUILLA BASKETRY

Basketry was such a critical industry in Cahuilla society that it is worth commenting on in some detail. Baskets were involved in almost every aspect of the daily collecting and food processing routine. They were usually made by young girls and very old women. When a girl was very young, she began learning how to make baskets from her grandmother. Women too aged to participate in the plant gathering process were still able to contribute significantly to the culture if they possessed the manual dexterity and keen eyesight necessary for basketmaking. A woman who was adept as a fine craftsman acquired considerable prestige within her community and had an economic advantage over others since she could trade her product for other commodities.

Cahuilla baskets are renowned for their closeness of weave, elaborate designs, and aesthetically pleasing interfacing of natural and dyed colors in design backgrounds. The design elements were never random, but were made up of symbols of significance to the culture or individual basketmaker. Cahuilla basket designs frequently represent spirit beings such as rattlesnake, eagle, or whirlwind. Certain figures associated with the Cahuilla religion (such as owls, coyotes, foxes, and bears) appear never to have been depicted on Cahuilla baskets, however, nor were symbols associated with death and danger. Often economically important plants were depicted on baskets in a highly stylized manner.

Baskets were made for ritual and ceremonial functions as well as for daily household and gathering use. Baskets were used to carry ritual tobacco during ceremonies. They were burned as mortuary offerings at funerals. In addition to ceremonial functions, they were presented as gifts to guests who participated in the many Cahuilla ceremonies. Such gifts testified to a deeply personal relationship between the giver and receiver, and were usually thereafter viewed as treasure items rather than utilitarian objects. Possession of gift-baskets was an indicator of status and wealth, and such baskets were always translatable into subsistence resources through exchange.

Baskets were maintained in good condition by frequent washing with saponaceous materials. They were repaired by their owners until they were no longer serviceable. Flaring baskets which had become worn in the center could be mounted with various pitches such as pine gum or asphaltum as rims to stone mortars. When a basketmaker died, the baskets that she owned were destroyed. Furthermore, specific idiosyncratic designs which had characterized the work of a basketmaker were discontinued after her death.

Cahuilla basket forms included shallow flaring bowls, large deep baskets, flat plates, small bottleneck-like receptacles, and a woman's cap. The large deep baskets served for carrying and storing plant foods. A number of plants were used to produce dyes for basketry designs, and the colors achieved are often extraordinary in their subtlety and gradation. A reddish color was often obtained by dipping the basketmaking materials into a rusted iron pot in the post-contact period. Another innovation was a black dye made by soaking basket rushes in a solution of walnut hulls. Mrs. Victoria Wierick said that the rushes had to be placed in the dying solution for a predetermined interval or the plant materials would rot. In the case of walnut hull dye, the rushes were kept in the dye solution for a period of two weeks to two months depending upon the shade of color desired. Coffee grounds were another source of black dye in recent times.

For many years, beginning in the late nineteenth century, Cahuilla women contributed significantly to the economy of their people by manufacturing baskets for commercial sale. The industry was widespread throughout southern California, and baskets from the Cahuilla and other tribes of the area were sold throughout the country to museums and collectors as far away as New York City and Philadelphia. This industry resulted in many changes in traditional size, shape, and design of Cahuilla baskets. Entrepreneurs such as George Wharton James and persons interested in Indian rights and welfare, such as Constance Goddard DuBois, publicized and found varying outlets for this cottage industry. New and more elaborate designs were developed. One well-known basket, for example, includes both an ostrich and mother-hubbard skirt design elements. Exotic types of baskets were created to satisfy the new market demand: fruit baskets for tables, baskets for cannisters, enormously large baskets, and baskets of miniature size, some as small as a thimble.

Today basketmaking among the Cahuilla is virtually extinct. Not only is it difficult to locate the necessary plant materials to make baskets, but the time involved in creating a really fine basket makes its cost prohibitive on today's market. The decline in basketmaking can also be traced to the fact that in more recent times Cahuilla women began working off their reservations as domestics or in the fruit industry, and there was no longer an opportunity to learn and perfect the art. At the same time, young girls sent away to schools of the Bureau of Indian Affairs were no longer home at the time when they would have customarily learned the craft. Although there are still a few Cahuilla women who are familiar with the more complex basketmaking methods, most women who possess any knowledge of basketry are familiar with only the most rudimentary techniques.

CAHUILLA PLANT USE TODAY

Many Cahuilla regret the loss of traditional subsistence patterns, particularly the mixed and well-balanced diet that traditional foods provided. Longevity, mental alertness, and good eyesight are associated in the minds of many Cahuilla today with their traditional diet. They believe that adopted foods have brought about a general physical weakness, shortened life-span, a tendency to obesity, and a proneness to such diseases as diabetes. Among women, dysmenorrhea, painful childbirth, and stillbirths are attributed to the new dietary habits.

Although some aboriginal foods are still collected and eaten, particularly at fiestas, decreasing availability of many plants because of environmental changes and the difficulty of gathering have contributed to a disappearance of most traditional foods. During the Depression, when the economy of the country so radically affected millions of people, Indians throughout southern California who still remembered the traditional plants were able to turn once again to their use. The knowledge of plant uses is fast fading, however, and it is unlikely that this will ever happen again.

Traditional plant medicines are still utilized to some extent. Some of the older cures are considered to be efficacious in many cases where frequent visits to a non-Indian doctor have failed.

Cahuilla remedies are used most frequently today in the healing of sores and swellings resulting from poor circulation of the blood, for ulcers and other intestinal disorders, and for colds and bronchial infections.

There can be little question that traditional Cahuilla foods provided well for these people. There were less fats and carbohydrates in the diet, only natural sugars, and food was generally fresh, retaining its nutritional elements. Processing methods employed by the Cahuilla rarely resulted in a deterioration in food value. Obesity was rarely caused by too much or the wrong type of food. Undoubtedly, the Cahuilla are correct in believing that many of the medical problems associated with modern Cahuilla life are related to changes in their diet.

While it is doubtful that there will ever be any substantial return to the traditional diet, there is greater interest today among Cahuilla young people and other American Indian people in the food plants of the past than at any other time in the past half-century. When Cahuilla speak of their grievances against the white-man, they frequently mention the loss of traditional foods. The fact that many ethnobotanical researchers are increasingly recognizing that the plant knowledge of the American Indian was highly sophisticated and that in recent years the positive contributions of Indians to contemporary medicine have been documented is a source of cultural pride to the Cahuilla as it is to other Indian people of America.

An Annotated List Of Cahuilla Plants

The plants used by the Cahuilla are presented in the annotated list that follows alphabetically by genera, followed by the species in alphabetical order within each genus. The family name is given in parenthesis after each scientific name. In addition, the common name of each plant has been supplied where available as well as every Cahuilla name that could be ascertained by the authors. A brief discussion of the use of each plant and sources of information follows. Where a Cahuilla plant name was not remembered or has not yet been determined, the abbreviation of N.D. (not determined) is employed. In a few instances, all members of a plant species used by the Cahuilla, such as oak and manzanita, are grouped together for the sake of a more integrated discussion in a single section under a genus heading or the common name (i.e., cactus).

In general, the plant names presented here are in accordance with those given by Philip A. Munz and David P. Keck in A California Flora (1959), where our readers may obtain more detailed information concerning the morphological characteristics of a species, its habitat, flowering dates, and altitudinal and geographic distribution. The discussions of plant usage are not intended to be definitive, but only to indicate the cultural context in which each plant was used.

The Cahuilla plant names and other words are presented here

in accordance with a practical spelling which has the following characteristics:

(1) The consonants of native Cahuilla words are *ch h k kh l ly m n ng ny p q r s sh t v w y*, plus the sound indicated here by an apostrophe. These should be pronounced much as in English, but with the following exceptions.

kh is the fricative sound of *j* in Spanish *José*, or *ch* in German *Bach*.

ly is always pronounced like the *lli* in English *million*.

ny is always pronounced as in canyon.

q is a sound which does not occur in English or any European language, although it occurs in Arabic. This is the sound we transcribe with *q* in words like *Iraq*.

The apostrophe is for a glottal stop—an abrupt interruption of air in the throat which resembles that between the two vowels of the English *oh-oh*.

(2) The Cahuilla vowels are *a e i u*, pronounced with their Spanish values, e.g., in padre, Jose, chico, luna. When written double, i.e., as *aa ee ii uu*, they should be held twice as long as the single vowels.

A technical description of Cahuilla sounds with notes on their derivation from the sound system of prehistoric Uto-Aztecan was presented by Bright (1965*a*). The lay reader who wishes to know more about Cahuilla will find a less technical guide to pronunciation in a field guide to southern California Indian languages published by Bright (1965*b*).

Abronia villosa Wats. (Nyctaginaceae)

Common Name: *Wild Sand-Verbena* Cahuilla Name: *temal nyuku*

The Cahuilla called the wild sand-verbena *temal nyuku* or "earth cousin," an appropriately poetic allusion to the prostrate growth characteristics of this plant which hugs the earth closely. A Cahuilla game was played with this plant by young children. Patencio (1943:59) described the game: "We children would pull up wild sand verbanas [sic] by the roots, and holding them above our heads, run around each other, and guess who was hiding in the vines." Mrs. Katherine Saubel identified another species, *A. umbellata*, as being used in such games.

Acacia Gregii Gray (Leguminosae)

Common Name: *Catclaw*   Cahuilla Name: *sichingily*

Catclaw ranges from four to eight feet in height and usually may be found in washes and on mesas of the Lower Sonoran zone, although sometimes it extends upward into the pinyon-juniper belt. Since catclaw grew in extensive groves in some areas, it probably provided a substantial, although not preferred, source of food for the Cahuilla. The authors have observed catclaw near Banning, on the western Colorado Desert, and in the Borrego Desert.

A low shrub, catclaw flowers from April until June. The bean-like pods begin appearing in May and usually last until August, depending upon the individual plant and its location. The pods, ranging from two and one-half to six inches in length, were eaten fresh or dried and ground into flour from which mush or cakes were prepared. Because the pods are sometimes bitter and slightly alkaloid, catclaw was not a favorite food. When the pods were too bitter, they were parboiled to remove the unpalatable taste. A seed analysis by Earle and Jones (192:230) showed a protein content of 33.8%, an oil content of 25.4%, and a trace of alkaloids.

Barrows (1900:60) reported that the pods produced food in sparse quantity and were used only occasionally. The fact that all Cahuilla interviewed by the authors immediately recalled this plant, however, and thought of it as a food source appears to testify to a much heavier and widespread usage throughout Cahuilla territory. The plant was also considered an outstanding construction material and a fine firewood.

Acer macrophyllum Pursh. (Aceraceae)

Common Name: *Big-Leaf Maple,*   Cahuilla Name: *sivily*
                Canyon Maple

The limbs of the big-leaf maple were used in house construction and considered good firewood.

Adenostoma fasciculatum H. & A. (Rosaceae)

Common Name: *Chamise, Greasewood*   Cahuilla Name: *u'ut*

The chamise, one of the most common plants in Cahuilla territory, was employed for several purposes. Abundant throughout chaparral regions and on dry slopes and ridges below 5,000 feet, the plant was available to most groups as a construction material. Barrows (1900:36) wrote: "The work of building is done by men. A quantity of stout poles six to ten feet long and from four to five inches in diameter are cut out from the 'greasewood.' Cahuilla *o-ot* (*Adenostoma fasciculatum*), which in these mountains attains the dimensions of a small tree." The branches were also employed in making arrows and building ramadas and fences. Cinciona Lubo recalled that her family used the large root, a foot or more in diameter, for firewood. The coals of this wood were a favorite source for roasting.

Chamise was traditionally employed as a medicine, and is used among the Cahuilla in this form today. Leaves and branches are boiled and the solution obtained is used to bathe infected, sore, or swollen areas of the body. According to Barrows (1900:79), an infusion of chamise was given orally to sick cattle in the late 1890's. Occasionally, branches of chamise were bound together for use as torches or part of a branch was used in conjunction with other wood to make bows.

A gum from a deposit of a scale insect of the chamise was used as an adhesive to bind arrow points to shafts, baskets to mortars, and for similar purposes.

Adenostoma sparsifolium Torr. (Rosaceae)

Common Name: *Ribbonwood,* Cahuilla Name: *henily*
Red Shank, Ribbon Bush

This familiar small tree of southern California's dry slopes provided a seasonal source of food, medicine, firewood, building materials, a wooden head for arrows, and more recently fencepost material. Ribbonwood is found throughout Cahuilla territory, particularly on chaparral hillsides below 6,000 feet. The seeds were ready for harvesting in July and August, and provided a limited but regular contribution to the total diet.

The tree's easily stripped off bark was used as a fibrous material for women's skirts. The tree's limbs provided a favorite firewood for roasting, giving forth a heat of high intensity. A more recent use of limbs has been in fencing agricultural fields.

Medicinal uses are best remembered today by Cahuilla. The leaves were used to make a beverage which relieved ulcers, and cured colds and chest ailments. Dried and steeped, the leaves produced a beverage said to aid stomach ailments by inducing bowel movement or vomiting. Ground into a powder and mixed with grease, the twigs provided a salve for sores (Barrows 1900:77). The late

Salvador Lopez, a curer of considerable reputation among the Cahuilla, used the plant for relief of arthritis. Alvino Siva recalled that the plant was mixed with bacon fat to make a poultice for saddle sores on horses.

Barrows (1900:50) recorded that ribbonwood was used as the head of two-piece arrow shafts. The main part of the shaft was made from a hollow reed (*Phragmites communis*) and the ribbonwood head was inserted into the hollow end of the shaft and fastened there with glue or by sinew wrapping The long, slender, flexible branches were also often employed as housing material or fashioned into throwing sticks, which were skillfully employed in hunting birds and rabbit. Although all Cahuilla questioned recalled the name of the plant as *henily*, the term *sankat* was recorded by Barrows (1900:77).

Agave deserti Engelm. (Agavaceae)

Common Name: *Agave, Century Plant*   Cahuilla Name: *amul*

Agave was a basic food staple of the Cahuilla, and as Aschmann (1959a:80) pointed out extremely valuable during drought periods. Since agave is available for harvesting in November and December, it was also an excellent food source during a rather lean time of the year for food plants. Among desert-oriented Cahuilla groups, agave occupied the important place in the food supply held by acorn and pinyon among other groups.

Where competition from other plants is not too great, agave makes up a large proportion of plant cover, but only a small proportion of agave in a given area sends up flower stalks in any one year. As a result, there was generally a constant supply of agave from year to year.

The agave commonly occurs throughout the western Colorado Desert, being especially abundant on the western and southern slopes of the Santa Rosa Mountains. Usually found in gravelly or rocky areas at from 500 to 3,500 feet, agave was available within the gathering boundaries of most Cahuilla landholding units. Cahuilla people interviewed by the authors have mentioned Cathedral Canyon, Palm Canyon, and the foothills above Palm Springs and the Anza-Borrego Desert as being particularly fruitful for gathering the plant. In Rockhouse Canyon and Coyote Canyon, both Cahuilla sib areas, agave grew abundantly.

The Cahuilla sibs in the desert gathered agave on the western slopes of the Santa Rosas. The Wanikik Cahuilla of San Gorgonio Pass gathered agave near Whitewater Canyon. Other specific agave gathering areas mentioned by Cahuilla were Rabbit Peak and areas just below Hidden Lake lying above Palm Springs. The sites of numerous agave roasting pits may be found also in the hills directly south of Palm Desert in the Deep Canyon region.

Agave gathering areas were owned by sibs and lineages, although ownership was not as clearly defined as with oak and mesquite stands, probably because the plant is destroyed in its gathering. As a result, one could not claim ownership of specific plants from year to year. Certain gathering areas, however, were claimed by specific lineages. When agave was ready for collecting, a group of men, usually representing a member of each family, traveled to their gathering area for the harvest. The gathering areas were usually five to ten miles from a village, and collection required several days. Although plants may be harvested in April and May, prime gathering times were in November and December. Agave gathering, primarily a masculine activity, was a festive time for telling stories, singing religious songs associated with agave as well as secular songs, and passing time in other entertaining ways while awaiting final harvesting of the plant.

Prior to the gathering, men were sent out to scout the best gathering places for that season. Since deer and mountain sheep often ate agave before it could be harvested, it was important that attention be paid throughout the season to areas where plants were likely to ripen.

During the harvest, young male children were taken along to learn the collection procedures. Juan Siva of the Isilsiva lineage was five or six years old when he was taken on his first gathering trip in about 1875. It was from Siva and his friend and age-mate, Calistro Tortes, that much of our data on agave gathering was acquired. In more recent times, gathering became a joint family affair with men, women, and children travelling in horse-drawn wagons to collection areas to gather and prepare the plants.

Agave as Food.—Three parts of the agave are edible: the flower, the leaves, and the highly prized stalk or basal rosette. The tenderness of the leaves is due to a rich supply of sap and water. The quality of the plant was partly related to rainfall; during dry seasons crops were less satisfactory.

Each agave produces several pounds of edible flowers between April and August. Aschmann (1957a:79) noted that starch stored in the buds is converted into sugar by a natural process, the sweet nectar being exuded by the flowers. Like the yucca blossom, agave flowers were parboiled to release their bitterness. After parboiling, they were eaten or preserved for future use by drying. When dried blossoms were needed, they were reboiled in water. Barrows (1900: 66) reported that the blossoms could be preserved for up to five years.

Agave leaves are edible at almost any period of plant growth. They are best from November through May when the plant is rich in sap. During the blossoming period, leaves were also cut and gathered. They were again gathered at the same time as the stalk or basal rosette (*yamal*). Slightly yellow leaves and those nearest the ground were bitter and therefore generally not collected. If yellow leaves were collected, they required special processing to reduce the bitterness. After gathering, the leaves were baked for immediate consumption or dried and stored for future use.

An historic incident attesting to the importance of agave leaves as a food source was recorded by José Maria Estudillo in 1824 (Bean and Mason 1962:47). The Cahuilla Indians near Palm Springs abruptly abandoned negotiations with a Mexican expedition through their territory so that the lineage leader and his men could go on an agave-gathering expedition with horses belonging to the Mexicans. The Cahuilla acted in this manner despite the fact that they had received valuable trade goods from the Mexicans and did not have permission to borrow the horses. The Cahuilla wanted these pack animals on which large quantities of agave could be transported. This is the first recorded use of horses as pack animals by the Cahuilla. The incident suggests a prior acquaintance with the animal.

An added delicacy of agave leaves was the frequent presence on them of the larvae of the Agave Skipper Butterfly (*Megathymus stephousi*). The larvae were roasted on the leaves and then picked off and eaten.

The favorite Cahuilla food from the agave was the stalk, which was ready at various times from April throughout the summer, prior to blossoming and after it had reached a height of about four to five feet. Only a certain number of edible stalks were harvested during each gathering, so that more would be available later in the season. Barrows (1900:5) described roasted agave stalks as sweet and tasting like molasses. The stalks weigh several pounds each, and could be gathered in abundance during a good season. Chase (1919:62) observed a Diegueño Indian gathering a dozen tender stalks in an hour or so. Thus, in a good collecting area, a group of men probably could gather several hundred pounds of stalks in a day.

The leaves of the plant were removed with a mescal cutter, a shovel-shaped instrument of hard wood with a sharp fire-hardened edge. The stalks were detached from the plant base with a sharply pointed hardwood pole, usually made of oak or ironwood. Although gathering destroyed individual plants, it did not interfere with plant reproduction since the agave sends out shoots from its base which take root nearby. This process occurs before the plant becomes edible. Occasional depletion of plants near villages may have been due to excessive use of agave for its food and fibrous qualities.

Food Preparation.—A pit about three feet deep and five feet long was dug by hand or with an agave shovel in sandy soil. A large rock was placed in the center of the pit and smaller rocks were placed around it. Logs were next placed in the pit and permitted to burn into a bed of long-lasting coals. The coals were covered with a layer of rocks, and agave stalks and leaves were laid across these rocks. The pit was then covered with grass and leaves to facilitate steaming and enhance the flavor of the roasted stalks. Several bushels of stalks and leaves could be roasted in one pit. The cooking process lasted three nights, each night having a special name. The first night was called *pemqualemu,* the second night *pasnas* (meaning bitter), and the third night *pasnavixniat* (meaning third night of cooking).

As women were judged by the skill with which they prepared foods, so were the men who prepared agave for roasting. The care which they took in cleaning and roasting agave was a matter of community concern and a means by which a man could be compared with other men. Thus, it was essential that young boys learn the correct techniques for agave preparation as soon in life as possible.

Preservation of Agave.—Agave flowers were dried for future use in the sun, and, as mentioned earlier, could be preserved for several years. They had to be thoroughly dried, however, or mildew became a problem. The mildew was referred to as *pisaqa* ("it's rotting"). Flowers usually were stored in dry caves so that dampness would not affect them.

Roasted stalks and leaves were pounded into cakes, dried in the sun, and stored in hermetically sealed pots. David Prescott Barrows in a lecture given in 1958 reported that he had saved an agave stalk for nearly sixty years and still found it edible. The fact that agave could be preserved so long as well as its natural abundance contributed to its role as a major Cahuilla food staple ranking with oak, mesquite, and pinyon.

Other Uses of Agave.—Although agave served primarily as a food, the plant had several other uses. The most important of these was as a fiber. Agave leaves provided a fiber ranking in strength with hemp and other modern commercial fibers. The leaves were pounded to release the fibers, which were then dried and separated by combing.

Agave fiber was used in making a wide variety of articles: nets slings, shoes, women's skirts, bowstrings, mats, cactus bags, cordage, cleaning brushes for cooking water, snares, and baby cradles. Women wet the fiber with saliva or water and twisted it on their thighs to form thread or string. Fiber nets, essential in Cahuilla hunting and gathering, were called *ikat* and were of the tump-line variety. The tensile strength of these nets was great and up to 100 pounds could be carried in them. Nets were used as baby cradles. Water ollas and other articles were hung in net frames from the rafters of Cahuilla homes.

Agave sandals were made by both men and women, who threaded numerous fibers into a soft cushion for the foot that was wrapped about the ankle with a fiber thong. The Cahuilla name for these sandals was *chawish*.

Agave netting was also used in ceremonial costumes. The eagle feather skirts worn by ceremonialists had a basic foundation of netting. Depending on the ultimate use intended, nets were made by either men or women. Women made nets for most secular uses, while the men made the netting used in ceremonial activities, which were exclusively their responsibility.

During the late nineteenth century and well into the twentieth century, the Cahuilla wove saddle blankets from agave fibers as a commercial product. These saddle blankets were highly prized by the Cahuilla and by the Mexicans and later Anglo-Americans. Photographs were made of the weaving process, but only one of these blankets exists in any museum collection (Smithsonian Institute).

Until recently, scholars have been puzzled about the origin of the agave fiber weaving industry among southern California Indians. The circumstances surrounding introduction of this industry, however, were recently related to the authors by Rosinda Nolasquez, a Cupeño. According to her, the industry originated with her great grandfather, a Yaqui Indian from Mexico who married a Cupeño woman. He taught the Cupeño at Warner Hot Springs a mechanized technique for producing agave fiber in quantity and weaving methods he had learned in northern Mexico. The saddle blankets (*kooku*) and other similar articles were then produced for sale and trade by the Cupeño people at Warner's. The weaving method gradually spread from Warner's to a number of surrounding reservations. After the dispossession of the Indians at Warner's and their forced removal to the reservation set aside for them at Pala in 1903, the industry was abandoned by the Cupeño.

According to Ruth Merrill (1923:235), agave leaf was used by both the Cahuilla and Diegueño as a foundation in making baskets. The agave leaf also provided a combined needle and thread which the Cahuilla used for sewing. The hard needle-like thorn at the end of each agave leaf will if carefully detached come out of the leaf with several feet of fiber attached, thus making a natural needle and thread combination.

The Cahuilla used agave thorns also for awls employed in basket-making. The thorns were set into wooden handles with asphaltum. In addition, agave thorns made useful tools for tattooing, a popular custom among the Cahuilla. Ashes of burned stalks were used as a dye for the tattoos. Dried stalks were utilized by the Cahuilla for firewood.

Allenrolfea occidentalis (Wats.) Kuntze (Chemopodiaceae)

Common Name: *Iodine Bush*   Cahuilla Name: *hu'at*

The small, black seeds of this sprawling shrub were ground and made into a mush or drink. Mrs. Alice Lopez recalled that the seeds tasted something like chocolate—the comparison having to do with the powdery consistency of the flour and not the taste. The powder was sprinkled into water to make the drink or mixed with a small amount of water to form mush. The flour also could be dampened, shaped, permitted to dry, and eaten as a cookie. The iodine bush is usually found growing in moist, alkaline soils, especially in desert areas. Mrs. Lopez remembered it as growing near the road to Torrez-Martinez reservation, where the soil is extremely dry and alkaline. Cinciona Lubo described a similar plant she called *toat*, which may be the same plant or simply reflect a Cahuilla term for flour or meal.

Allium biceptrum Wats. (Amaryllidaceae)

Common Name: *Wild Onion*   Cahuilla Name: *N.D.*

Romero (1954:62) reported an extract of this plant was combined with powdered berries of the *Rhus trilobata* to cure appetite loss. Since this species has not been reported in southern California, it seems likely that another *Allium* species was used for this purpoe.

Allium validum Wats. (Amaryllidaceae)

Common Name: *Wild Onion*   Cahuillah Name: *tepish*

A favorite delicacy of the Cahuilla, this purple-flowered wild onion may be found in the lower valleys, such as Tripp Valley and San Jacinto Valley. The plant was gathered in late spring or early summer, preferably just prior to and during the early period of flowering. Bulbs were eaten raw or used as a flavoring ingredient for other foods.

Amaranthus fimbriatus (Torr.) Benth. (Amaranthaceae)

Common Name: *Pigweed*   Cahuilla Name: *pekat*

This annual, found in the interior valleys and the desert areas of southern California, blooms and seeds if moisture conditions permit soon after the summer rains. Although not as dependable as some plants, pigweed was sought out by the Cahuilla, who made a mush from the seeds. Gathered in late summer, the small seeds were left on the spikes until needed. They were then threshed, parched, ground into a flour, and made into mush. Leaves of young plants were boiled and eaten as greens or used as potherbs.

Ambrosia psilostachya DC. (Compositae)

Common Name: *Western Ragweed*   Cahuilla Name: *N.D.*

Hall (1902:123) reported that western ragweed was common in the San Jacinto Mountains, where it was called *yerba sapo* or toad plant by the Indians. He specifically noted that it was a "wayside weed in Strawberry Valley." The plant has not been identified by present-day Cahuilla.

Amelanchier pallida Greene (Rosaceae)

Common Name: *Service-Berry*   Cahuilla Name: *N.D.*

This berry-producing shrub is scattered throughout Cahuilla territory, especially in alluvial bottom lands, moist valleys, and on mountain slopes. It is found frequently in association with manzanita and other chaparral-associated plants. The berry ripens from June until August at different locales. Eaten fresh, the berry provided a welcome addition to the summer diet. The berry was also preserved by drying and could be eaten several months after gathering. Zigmond (1941:191) stated that the service-berry compares favorably with cultivated fruits as a vitamin source.

Oscar Clarke has observed species of *Amelanchier* growing on canyon walls in the Santa Rosa Mountains, above Pinyon Flats campground, near Toro Canyon, and on the Pipes Canyon Road in the San Bernardino Mountains. The spotty occurrence of this plant may explain why it has not been recorded previously as used by the Cahuilla. Fruit appears to sit well during wet years only, and even then does not develop the juicy quality so well known in northwestern species.

Anemopsis californica Hook. (Saururaceae)

Common Name: *Yerba Mansa*   Cahuilla Name: *chivnish*

This perennial herb, common in wet, alkaline soils below 6,500 feet, was used by the Cahuilla as a cure for a variety of ailments. The strongly aromatic and peppery roots, according to Alice Lopez, were peeled, cut up, squeezed, and boiled into a decoction that was drunk as a cure for pleurisy. Mrs. Lopez also recalled use of the infusion as a cure for stomach ulcers, chest congestion, and colds. Katherine Saubel uses an infusion made from the bark of the plant to wash open sores. The bark was gathered in the fall, according to Mrs. Saubel, boiled into a deep red-wine color, and then drunk as a cure for ulcers or applied externally to sores as a wash. In earlier times, collecting was usually carried out by men, since the plants grow high up on hillsides and were usually some distance from villages.

Romero (1954:15) reported that the plant was "cut, dried, and powdered and used for disinfection of knifecut wounds, and to draw and promote the growth of healthy flesh." Palmer (1878:650) noted: "The root of this plant is a great remedy among the Indians of southern California. It has a strong peppery taste and odor.

A tea made from the roots and a powder prepared from the same and applied to venereal sores are a great remedy. The powder . . . [is] . . . used on cuts and sores, as it is very stringent. The leaves after being wilted by heat and applied to a swelling are a sure cure."

Barrows (1900:78) may have been referring to this plant when he noted that the "root of an unidentified plant, called *chi-vi-ni-vish* is cleaned from just where it branches into the stem, pounded up, boiled into a dark draught, and used to alleviate any sort of pain." Calistro Tortes stated that an infusion from this plant was used to cure open sores on cattle.

Apiastrum angustifolium Nutt. (Umbelliferae)

Common Name: *Wild Celery*   Cahuilla Name: *pa'kily*

The wild celery, a small, fragile hairlike plant which bears no resemblance to celery, provided a small seasonal food source in wet years. The plant is found growing in loose soils and sands. Bowers (1888:6) first reported this plant in use among the desert Cahuilla.

Apium graveolum L. (Umbelliferae)

Common Name: *Common Celery*   Cahuilla Name: *pa'kily*

Common celery, introduced into America from the Old World, was widely adopted as a potherb. The plant abounds in wet areas, particularly during May to June. Romero (1954:68) first noted a Cahuilla medicinal usage. A decoction was employed for ailments attributed to kidney malfunction. Romero also recorded the term *"se-ma-mek"* as the Cahuilla name for this plant, although the term is no longer recognized. Since *pa'kily* is also the Cahuilla word for wild celery, the term may be a general name for celery.

Apocynum cannabinum L. var glaberrimum A. DC. (Apocynaceae)

Common Name: *Indian Hemp*   Cahuilla Name: *N.D.*

Throughout the Southwest, this plant was widely recognized for its medicinal properties and usefulness as a fibrous material. Found in damp soils below 500 feet, the plant is considered toxic by many authorities and has been said to poison livestock, although no cases of human poisoning are known (Hardin and Arena 1969: 106). Palmer (1878:755) wrote of its use in Arizona as follows:

"Indian hemp or silk, as it is sometimes called, is very extensively used by the Indians of Arizona for the manufacture of twine and cloth. The bark of the plant is tough and strong and sometimes like flax. The Indians cut the plant when ripe and rub it so as to separate the fibers, with which they make very strong and beautiful fishing lines, and a fine thread which they use in sewing and also make into cloth. The root gives out a bitter milky fluid that is used as a medicine by the Indians." In another passage, Palmer (1878:649) noted that the Indians in California also "use the fibre prepared from the stems of this plant to make rope, twine and nets." Drucker (1937:21) confirmed the use of Indian hemp for fiber by the Cahuilla.

To remove the fiber, the woody stems were first soaked in water. The bast with the back could then be removed easily. The latter after washing left a soft, silky fiber of yellowish-brown color which was strong and durable. Palmer (1878:649) wrote: "I have seen ropes made of it that have been in constant use for years."

Arbutus Menziesii Pursh (Ericaceae)

Common Name: *Madroño, Madrone*   Cahuilla Name: N.D.

Mrs. Alice Lopez recalled that leaves of this plant were used to make a medicine for stomach ailments.

Arctostaphylos Adans. (Ericaceae)

Common Name: *Manzanita*   Cahuilla Name: *kelel*

Five species of manzanita, two of which have not been determined, were used by the Cahuilla. The three remembered species were A. *glauca* Lindl., A. *pungens* HBK., and A. *glandulosa* East. All five species were referred to as *kelel*.

Manzanita is most commonly found in chaparral and oak-chapparal regions and may be seen in profusion along the road from Banning to Idyllwild in the San Jacinto Mountains, where one may form an excellent idea of its distributional characteristics. It is commonly found also in the lower canyons where Cahuilla village sites were once located.

Manzanita as Food.—The light-brown to red-brown fruit of manzanita customarily was gathered from June until September. Early in the season, when the berries were slightly red, they were considered suitable for use in a beverage. The pulp was mashed, mixed with water, and strained into a drink. A simpler method of making this same drink was to soak berries in water without crushing them. When fully ripened, manzanita berries were used in making a gelatinous substance that was eaten like an aspic. The berries were also sun dried and stored in ollas for future use.

Calistro Tortes recalled that the berries became bitter after drying, although they were still used and ground into flour from which a mush was made. The lack of sweetness was compensated for by adding wild honey or mixing the flour with other more palatable flours. Although berries of all manzanita species were eaten when dried as a food, the Cahuilla preferred A. *glandulosa* and A. *pungens* as fresh fruit, since the berries of A. *glauca* were said to be too sticky when fresh.

Manzanita seeds were ground into a meal from which a mush or cakes were prepared. Leaves were occasionally mixed with tobacco or steeped in water to make a tea used in curing diarrhea or poison oak rash. Krochmal, Parr, and Duisberg (1954:8) noted that in Arizona the leaves were steeped in water and given as a remedy for stomach ailments.

Although Barrows (1900:64) reported that the Cahuilla name for manzanita was *tatuka*, present-day Cahuilla do not recall this name.

Manzanita was a primary food source of the Cahuilla, since it could be gathered in large quantities. Like other berries, it was important to collect the fruit as soon as it ripened since birds could deplete the supply rapidly. Tracts of manzanita were owned by lineages, and families had long-term use rights to specific tracts of manzanita plants.

Other Uses of Manzanita.—Manzanita wood was a preferred firewood that provided a hot fire and long-lasting coals. Branches were often employed in house construction. Barrows (1900:43) reported that awl handles were made from manzanita wood. Occasionally, stems of manzanita were used to make pipes and other small tools.

Manzanita also served as an indicator of wild game, since many small game animals, including deer and mountain sheep, ate the berries. Squirrels, chipmunks, kangaroo rats, and numerous birds clustered in manzanita thickets during the summer, providing a rich hunting opportunity for the Cahuilla while gathering food plants.

Artemisia californica Less. (Compositae)

Common Name: *California Sagebrush*   Cahuilla Name: *hulvel*

California sagebrush is considered one of the most important medicinal plants used among the southern California Indians, and it had a variety of uses. In particular, it was considered an essential plant to the proper maturation of girls into womanhood. Its primary use was to induce menstrual activity and assure a comfortable childbirth and rapid post-natal recovery.

Beginning with the onset of menstruation, young girls were given a tea made of the boiled plant and provided elaborate instructions concerning womanly arts. The decoction was taken just before the commencement of each menstrual period throughout a woman's life. This use was accompanied by various menstrual restrictions: for example, no salt, grease, or meat could be eaten for several days after drinking of the tea. Cahuilla women maintain dysmenorrhea was rare because of such dietary restrictions, and that drinking of *hulvel* alleviated menopause trauma. The drink was given to newborn babies one day after birth to flush out their system.

Leaves of California sagebrush were also used to relieve colds. The leaves were chewed fresh or dried and smoked after mixing with tobacco and other leaves. The plant was also used in Cahuilla sweathouses for various cures.

Artemisia ludoviciana Nutt. (Compositae)

Common Name: *Sagebrush, Wormwood*   Cahuilla Name: N.D.

This perennial plant grows in dry open places throughout Cahuilla territory at elevations below 5,000 feet. Barrows (1900) appears to have confused the plant with *Pluchea sericea* or arrowweed, thereby leaving open to question some of the uses he reported.

According to Barrows (1900:38), the plant was a favorite material in the Coachella Valley for roofing houses and in the wattling of walls. Barrows (1900:52) said granaries in the same region were almost exclusively made from this plant. A third use recorded by Barrows (1900:50) was employment of the shoots as arrow shafts. Kroeber (1925:704) also reported the Cahuilla used *Artemisia* for arrow shafts, but failed to designate the species. Probably several species of *Artemisia* were used in making arrows. The Luiseño regularly made their inferior arrows from *Artemisia Douglasiana* Bess. (Kroeber 1925:650), which may also be found in Cahuilla territory. *Artemisia dracunculus* L. would also have provided shaft material. Merrill (1923:236) noted that stems of the sagebrush were also utilized for both the warp and woof in Cahuilla basketry.

Artemisia tridentata Nutt. (Compositae)

Common Name: *Basin Sagebrush*   Cahuilla Name: *wikwat*

This pleasantly aromatic sagebrush may be found throughout Cahuilla territory from about 1,500 feet in the lower desert through the pinyon-juniper regions, extending in some areas as high as 10,000 feet or more.

The seeds of the basin sagebrush, which ripen in summer and fall (August through October), were gathered by Cahuilla women in large quantities. They were parched and then ground into flour for a pinole-like mush (Barrows 1900:65).

Leaves of basin sagebrush were made into a medicinal tea, which was administered for stomach complaints (Barrows 1900:78). The leaves and stems of the plant were also used as an air purifier or disinfectant. The dried leaves and stems were burned for this purpose in Cahuilla homes and in sweathouses.

The shoots of the plant were sometimes used in house construction, being laid across the rafters for roofing material or used in construction of the walls.

Asclepias L. (Asclepiadaceae)

Common Name: *Milkweed*   Cahuilla Name: *kivat or kiyal*

At least three species of this genus, *A. erosa* Torr., *A. fascicularis* Decne., and *A. eriocarpa* Benth., were gathered by the Cahuilla for use as gum, fiber, and possibly food. The plant is found on dry slopes and washes below the 7,000 foot elevation.

Hardin and Arena (1969:109) report that *Asclepias* species are highly toxic and extremely poisonous if eaten. Possibly one of the species lacks the toxicity attributed to the genus, since Cahuilla recall a milkweed plant that provided greens from May until June, when the leaves were gathered and parboiled. Later, the seeds of this same plant were ground into flour. Barrows (1900:75) reported that the white sap of *Asclepias*, which exudes when the plant is cut, was collected in a cup and set aside one night to solidify.

"Sometimes it is heated over the fire," Barrows noted, "After being thus coagulated, it is chewed. Its bitter taste disappears after a little, and a tasteless gum is left which affords inexhaustible satisfaction to the user."

Stems of the plant furnished a sturdy, fibrous material for nets, slings, and snares to capture small game. The stem was pounded to loosen the fiber, which then was extracted by rubbing the stem between the palms of the hands. The fiber was rolled on the thigh to produce cordage, whose usages testified to its natural strength and durability.

A fourth species of the genus, *A. syriaca* L., was said by Romero (1954:51) to have been used by the Cahuilla as an adhesive, although the authors have been unable to find any present-day Indian people familiar with this species. Romero gave the Cahuilla name for this plant as *semat hap-pac*. Cinciona Lubo suggested this may have been a rendition of the words *samat hepi'* or "bush milk." She recalled that the gum of a plant called *samat hepi'* was used to relieve pain from insect stings (ants in particular), the gum being applied to the injured area.

Astragalus L. (Leguminosae)

Common Name: *Locoweed, Milkvetch,*   Cahuilla Name: *qashil*
                   Rattleweed

Many species of this genus are notorious stock-poisons, and all species should be suspected of being toxic unless proven otherwise. Barrows (1900:66-67) reported the use of one species of locoweed, but was not able to botanically identify it. He wrote as follows: "One such species, called by the Coahuillas *kash-lem*, has a curious use as a flavoring principle. By summer-time the leaves fall away from the sere and yellow branches of these plants, and they are covered by quantities of straw-colored pods as big as the joint of a man's thumb. These quiver and rattle with every motion of the air, and give it one of the designations by which it is known. These pods, according to Celestino Tortes, are pounded up and mixed with beans, and perhaps other articles of food, as a spice."

Cinciona Lubo recalled such a plant, which she said was referred to in English as "sweet pea." She remembered that seeds were available in late summer, and said they were used as reported by Barrows.

Atriplex L. (Chenopodiaceae)

Common Name: *Saltbush*   Cahuilla Name: *N.D.*

At least two members of this genus, A. *lentiformis* (Torr.) Wats., a native, and A. *semibaccata* R. Br., an Australian introduction, were used by the Cahuilla.

A. *lentiformis* is found frequently in the lower and upper Sonoran life zones, and the authors have observed it from Banning to Anza in many locations. The seeds can be harvested from July to September, and were collected with a seedbeater and gathering basket. The seeds contain 13.6% ash, 12.5% protein, and 2.1% oil (Earle and Jones 1962:227). The seeds are somewhat larger than those of the popular chia (*Salvia columbariae*), and were prepared by parching, grinding them into flour, and mixing the flour with water to make mush or small cakes. This food was stored in great abundance, and in more recent times often mixed with wheat and corn mush.

Leaves of the plant contain some saponin and were used as a soap. The leaves and roots were crushed and rubbed into articles requiring cleaning. In addition, the plant had medicinal uses. Flowers, stems, and leaves were crushed, steamed, and inhaled for the relief of nasal congestion. Fresh leaves were chewed to relieve head colds. Dried leaves were smoked for the same purpose.

Mrs. Alice Lopez said that a "wormlike thing" (quite probably a beetle larvae) was gathered when the plant bloomed. It was then roasted and was "good to eat like popcorn." Another insect found on the saltbush was the cicada (*Diceropocta apache*). The cicadas were gathered in large quantities and roasted and eaten.

An introduced plant, A. *semibaccata*, commonly called the Australian saltbush, produces a small sweet and salty berry, which Mrs. Lopez reported was collected and eaten fresh by Cahuilla living in the vicinity of Indio.

A third saltbush species, A. *californica* Moq., which usually grows along the coast, was reported as having several uses by Palmer (1878), while he was working among the Indians of the Colorado Desert. Palmer (1878:603) said that the long roots of the plant were used by both Mexicans and Indians as a substitute for soap, after being pounded and mixed with water. He added: "It is said to be especially good in cleaning woolen fabrics. The seeds of this plant are also gathered, parched, reduced to flour, and made into mush or bread. At other times, the seeds are ground without parching and used as if parched."

Avena fatua L. (Gramineae)

Common Name: *Wild Oat*   Cahuilla Name: *N.D.*

This introduced plant, which commonly grows in the lower and upper Sonoran life zones, has long been used by Indians of the Southwest and was abundant in California as early as 1835 (Saunders 1914:53). The seeds were gathered from July through September, parched, ground into flour, and mixed with other wild seeds in mush. Wild oats are still used as a breakfast food by some Cahuilla and are believed to contribute to high energy.

The only name Cahuillas recalled for this plant is the Spanish *avena*. A highly nutritious seed plant, Earle and Jones (1962:224) reported that wild oat seeds have a protein content of 14.8%, an oil content of 11.5%, and an ash content of 2.8%. They also contain some starch, but no tannin or alkaloids.

Another Old World grass, *Avena barbata* Brot., was also gathered by the Cahuilla, who used its small edible seeds for food.

Baccharis viminea Pers. (Compositae)

Common Name: *Mule-Fat, Water-Wally,*   Cahuilla Name: *paq'ily*
                         Seep Willow

A common tree along water-courses and in wet areas, the mule-fat, also known as *yerbo de pasmo,* had four primary uses. Barrows (1900:78) reported that the leaves were steeped and used as an eyewash. Mrs. Katherine Saubel recalled that the leaves were considered a preventative for baldness. At the onset of baldness, hair was washed in a solution made from the leaves to prevent further loss. Mrs. Alice Lopez stated that the leaves and stems were boiled into a decoction used as a female hygienic agent. She also said limbs and branches were a favorite material used in house construction.

Baeria chrysostoma F. & M. (Compositae)

Common Name: *Goldfields*   Cahuilla Name: *aklakul*

Edible seeds of this plant were collected in June, parched, ground into flour, and used to make mush. The plant is small, yellow, and resembles a daisy. During the spring, goldfields grow in vast numbers in open grassland up to the 3,000 foot level.

Beloperone californica Benth. (Acanthaceae)

Common Name: *Chuparosa*   Cahuilla Name: *pisily*

This low shrub with its attractive flowers with scarlet corolla was called *pisily* by the Cahuilla, a word meaning sweet. The Diegueño are known to have sucked the flower for its nectar. The Cahuilla name for the plant suggests they were familiar with this usage.

Bloomeria crocea (Torr.) Cov. (Amaryllidaceae)

Common Name: *Golden Stars*   Cahuilla Name: *N.D.*

This plant, commonly found in the San Jacinto Mountains and on dry flats and hillsides up to 5,000 feet, has an edible corm. The corm was eaten raw at almost any time of year.

Brassica geniculata (Desf.) J. Ball. (Cruciferae)

Common Name: *Mustard*   Cahuilla Name: *N.D.*

A European introduction, the leaves of the mustard provided a food similar to the mustard greens found in today's markets. Although somewhat fuzzy in texture, the leaves are tasty and were eaten either fresh or boiled. The plants grow all winter where there is sufficient moisture and must be eaten in fresh condition. Mrs. Alice Lopez recalled that mustard was an important winter food plant. The seeds were also collected and ground into mush.

Brodiaea Sm. (Amaryllidaceae)

Common Name: *Wild Hyacinth*   Cahuilla Name: *mehawot* [?]

A number of species of this genus, among which *B. pulchella* (Salisb.) Greene is dominant, are common to the lower and upper Sonoran zones of Cahuilla territory. Munz (1965:1380) reports that the corms of a number of species were eaten by California Indians, and notes that the wild hyacinth has "horticultural merit and possibility." The corms (often called "bulbs" by the layman) were highly prized by the Cahuilla as food.

Collection of the corms could be made in great abundance as early as February on the lower desert and as late as July in the higher pinyon-juniper areas. A digging stick was employed in gathering. Mrs. Jane Penn remembers that only the large corms were kept as food when she was a girl. The smaller corms were replanted to ensure a crop the following year. The corm may be eaten either raw or cooked. When cooked, it was usually boiled for about a half hour. Romero (1954:40) stated that the corms and flowers were used as a soap and shampoo. He gave the Cahuilla name as *mehawot*, but the authors have been unable to obtain confirmation of this name.

Bromus tectorum L. (Gramineae)

Common Name: *Cheat Grass, Downy Cheat*   Cahuilla Name: N.D.

A native of Europe, cheat grass was considered a famine food by the Cahuilla. Seeds were gathered in quantity during periods of food shortage and cooked into a gruel, apparently without grinding, since the seed is very small. The plant, dominant in grassy areas of the San Jacinto Mountains and other mountainous areas, is not well-remembered by the Cahuillas. It was identified and its usage pointed out by Calistro Tortes to Harry C. James. Undoubtedly, many other grasses in Cahuilla territory were similarly used during periods of food shortage.

Bursera microphylla Gray (Burseraceae)

Common Name: *Elephant Tree,*   Cahuilla Name: *kelawat eneneka*
             Elephant Trunk

The elephant tree is found between Fish Creek and Carrizo Creek on the western Colorado Desert and from Arizona to Lower California. The plant was described rather than identified directly in the field by Cahuilla. Mrs. Alice Lopez recalled seeing it only once and in the area of the Santa Rosa Mountains. The name given for the plant is a descriptive word meaning "bitter wood" rather than the Cahuilla taxonomic name, which was not remembered.

Mrs. Lopez recalled that "great power" was associated with the elephant tree, which exudes a red sap when cut that she referred to as "blood." She specifically remembered that the sap was used in a cure for skin diseases, but she believed that the plant could be used as a cure against almost any disease.

Because the plant was associated with a great amount of "power," it was usually administered by shamans (*puvulam*). Players of peon, a popular gambling game among the Cahuilla, also used the sap to acquire "power" in their games. Mrs. Lopez's stepfather, who used the plant for curing, rubbed sap on the body of a patient. The sap was considered too dangerous to keep openly around a household, and was always hidden.

Unknown Cactus Species (Cactaceae)

Common Name: *Cactus*   Cahuilla Name: *navtem*

Throughout the spring and summer, various cactus species so characteristic of the southern California desert areas provided the Cahuilla with an important and dependable food source. Nearly all of the cacti provide edible berries or stems, and these foods were highly favored by the Cahuilla, ranking in importance with acorns, mesquite, agave, and yucca. The Cahuilla recognized and used more than a dozen different species of cactus, all of which they included under the generic name of *navtem*. Among species which the authors have identified as used by the Cahuilla are: *Echinocactus acanthodes* Lem., *E. polycephalus* Engelm. & Bigel., *Opuntia megacantha* Salm-Dyck., *O. Bigelovii* Engelm., *O. acanthocarpa* Engelm. & Bigel., *O. occidentalis* Engelm. & Bigel. var *megacarpa* (Griffiths) Munz., *O. ramosissima* Engelm., and *O. Parryi* Engelm. These species are discussed as they appear alphabetically in this work.

Several cacti were mentioned by Cahuilla or are referred to in the literature which have not been identified scientifically. These are discussed below under their Cahuilla names:

Chungal.—This term may refer to any plant which is thorny, but it was also applied to a specific cactus. Varying reports concerning its use were elicited from Cahuilla interviewed. One Cahuilla used the name *chungal* in reference to *O. ramosissima;* another said that the correct name for a plant referred to as *chungal* was *navet pengki* ("like cactus"), and that it was used as a medicine for venereal diseases. A third person said *chungal* referred to a specific cactus which was not eaten.

Mrs. Alice Lopez recalled *chungal* as a cactus commonly used by people living in the desert. She said that the buds were cooked and eaten just after the plant flowered, "when they are dry." The stems, she added, were picked when immature and cooked or roasted in a pit. Sometimes the spines were removed and the stems were boiled.

Chaal.—Juan Siva recalled a cactus called *chaal* with edible buds. The buds were knocked off with a stick, placed in a bag, and the bag was shaken to remove the thorns. The buds were then boiled or roasted.

Tinupem.—Barrows (1900:68) spoke of a cactus found "along the slopes of Torres mountain," which "might easily be mistaken for a neglected and stunted growth of the cultivated tuna." He described it as yielding "luscious fruit in large quantities."

Navtem.—Growing in the same vicinity as the *tinupem,* according to Barrows (1900:68), was a less thrifty cactus that grew close to the ground. The *navtem* was described as having flat stems and long spines of up to two or three inches. The fruit was comparable in yield and quality to the *tinupem.*

Ayuvivi.—Barrows (1900:68) noted that this unidentified species was very small, measuring "only about four inches high and covered with little hooked spines." The fruit was small and sparse.

Calochortus Pursh. (Liliaceae)

Common Name: *Mariposa Lily,*   Cahuilla Name: *talyki'* [?]
Star-Tulip, Butterfly Tulip

Several species of this genus, among them *C. catalinae* Wats., *C. flexuosus* Wats., *C. palmeri* Wats., and *C. concolor* (Baker) Purdy, may be gathered from May until August in the lower Sonoran and arid parts of the upper Sonoran zones of Cahuilla territory. Their highly nutritous bulbs, smaller than a walnut, were roasted in hot ashes in pits or steamed prior to eating.

The Cahuilla name for these plants could not be specifically ascertained, although there is a possibility that *talyki'* or "Indian potatoes" may have been the term applied to the mariposa lily. C. F. Saunders (1914:133) provided an excellent description of how the bulbs were processed. They could be eaten raw or after cooking. He described the method of cooking as follows: "A pit would be dug in the ground and lined with stones. Into this a quantity of fire-wood would be placed and ignited, making a huge bonfire which would heat the stones, and upon dying down would leave a good bed of ashes. Upon these 'the potatoes' would be spread, covered with a thick layer of leaves or brush and upon this would be laid a covering of dirt sufficient to imprison all the heat. On top of all another fire might or might not be built, and then the cooks went about their business. After the expiration of a given time, perhaps twelve or twenty-four hours, the pit would be opened and the bulbs taken out ready for consumption."

Capsella Bursa-pastoris (L.) Medic. (Cruciferae)

Common Name: *Shepherd's Purse*   Cahuilla Name: *N.D.*

This annual herb, an Old World native, is found throughout the year in Cahuilla territory from the desert floor up to 7,000 feet. The Cahuilla gathered it from January to June for use as greens. The edible seeds were collected from June to August. According to Romero (1957:7), the plant was used to make a tea that cured dysentery. He warned against drinking more than two cups.

Castilleja foliolosa H. & A. (Scrophulariaceae)

Common Name: *Indian Paint-Brush*   Cahuilla Name: *N.D.*

A root parasite, the Indian paint-brush is found below 5,000 feet in dry, rocky places. Mrs. Alice Lopez remembered only that this plant was picked by children who sucked the nectar from the flower.

Ceanothus L. (Rhamnaceae)

Common Name: *California Lilac,*   Cahuilla Name: *iswish* [?]
              Wild Lilac

Several species of California lilac are found in Cahuilla territory. Romero (1954:66) wrote that the California lilac had the ability to protect Indians when storms were at their worst near the timberline on high mountains by deflecting lightning. He recorded the Cahuilla name as *"o-oot,"* which may be a confusion with *u'ut* (*Adenostoma fasciculatum*). Cinciona Lubo gave the name of the "wild lilac" as *iswish*, possibly a generic name. Cahuilla interviewed by the authors did not recall any protective virtue attributed to the plant, but said it was used as firewood.

Centaurium venustum (Gray) Rob. (Gentianaceae)

Common Name: *Centaury, Canchalagua*   Cahuilla Name: *N.D.*

Palmer (1878:652) observed this plant being used as a common remedy for ague by Indians in southern California. He said a tea was prepared from the plant that served as a "very good substitute for quinine." The neighboring Luiseño called the plant *ashoshkit* and used it to reduce fever (Sparkman 1908:230).

Cercidium floridum Benth. (Leguminosae)

Common Name: *Palo Verde*   Cahuilla Name: *u'uwet*

The palo verde, a diagnostic plant of the Lower Sonoran zone, commonly is found in canyons of the western Colorado Desert and in scattered stands on the desert. The tree produces a slender, flat bean of about two to three inches in length. Beans were picked from July to August, dried, ground in mortars into flour, and used in mush or cakes. The trees are often large enough to shelter campers. Barrows (1900:60) recorded the name as o-o-*wit*, which essentially coincides with Mrs. Saubel's *u'uwet* as used among the mountain Cahuilla. Hansjakob Seiler (private communication to the authors) recorded the word *ankichem* for the plant among Cahuilla of the desert.

Chaenactis glabriuscula DC. (Compositae)

Common Name: *Pin Cushion*   Cahuilla Name: *N.D.*

Seeds of this winter annual were gathered from June until August. They were parched, ground into flour, and mixed with other seeds to form a mush. The plant is found in sandy valleys and foothills from the Lower Sonoran to Upper Sonoran life zones. Cinciona Lubo said that the seeds had a strong taste.

Chaetopappa aurea (Nutt.) Keck. (Compositae)

Common Name: *None in Munz*   Cahuilla Name: *N.D.*

This slender annual may be found in dry open areas and on grassy slopes up to 6,000 feet. Alice Lopez remembered the use of the plant pollen as a cosmetic for women.

Chenopodium L. (Chenopodiaceae)

Common Name: *Goosefoot, Pigweed*   Cahuilla Name: *ki'awet*

Several species of goosefoot are found in Cahuilla territory, including *C. californicum* (Wats.) Wats., *C. humile* Hook., *C. Fremontii* Wats., and *C. murale* L. A number of these species provided shoots and leaves which could be boiled and eaten as greens.

Barrows (1900:57) wrote in particular of *C. Fremontii*, which he said was gathered in large quantities and ground into flour for cakes. He added: "After a good harvest of this Chenopodium the edge of the Coyote canyon will be fringed with granaries holding stores of this food."

Barrows said the Indian name for the plant was *kiet*, which is probably the same as *ki'awet* above, which is the name Mrs. Katherine Saubel used for *Chenopodium californicum*. The latter species, common on dry slopes and plains below 5,000 feet was used as a food, soap, and medicine. Seeds were parched and ground into flour. Its carrot-like, hard root was also stored until needed, and then grated on a rock to make soap (Barrows 1900:48). Material requiring washing was rubbed in water with the scrapings. Although the leaves were not as efficient, they could also be employed as soap. Mrs. Saubel recalled that her mother sometimes boiled an entire plant and made a decoction for the relief of stomach disorders.

Cahuilla reported that a gum could be made from the milky sap in the stem of the goosefoot. The sap was also used to make a strong anti-helmenthic (Krochmal, Parr, and Duisberg 1954:8). An analysis of seeds of *C. fremontii* showed that they contain 12 to 17% protein, 1 to 4.6% ash, and 7.3 to 27.7% oil.

Chilopsis linearis (Cav.) Sweet (Bignoniaceae)

Common Name: *Desert Willow*   Cahuilla Name: *qaankish*

The desert willow, found throughout Cahuilla territory along desert washes and watercourses below 5,000 feet, is especially abundant on desert borders near mountain slopes. The tree produces both blossoms and narrow seed pods (six to eight inches long), which are edible, although they were not a major food source of the Cahuilla.

The desert willow was sought primarily as a source of wood. It was considered ideal for house frames, since the wood had great longevity and was also very pliable, a necessity for constructing the various styles of houses built by the Cahuilla. The wood was also employed in making granaries for the storage of mesquite, acorn, and other foods.

The strength and pliability of the wood made it an attractive material for use in bowmaking. Bowers (1886:6) reported that the wood was also used to hold ollas. He described this usage as follows: "They take the forks of the willow or some other trees with prongs starting out in several directions and dress them so that the ollas fit snugly within and then place them in and around their dwellings."

The long limbs of the desert willow were also used as sticks to reach plant fruits and nuts that were too high to grasp by hand. The bark provided a fibrous material useful in making nets, shirts, and breechclouts. The tree itself was usually comfortable to camp under, providing some shade for the desert dweller.

Chlorogalum pomeridianum (DC.) Kunth. (Liliaceae)

Common Name: *Soap Plant, Amole*   Cahuilla Name: *mocee* [?]

The soap plant may be found scattered throughout Cahuilla territory on dry open hills and plains, usually below 5,000 feet. The bulb is rich in saponin, which provided soap for the Cahuilla. The bulb was crushed and the particles rubbed in water to produce lather.

The coarse husk of fibers surrounding the bulb were removed and tied together to produce a sturdy cleaning brush that resembled a whisk-broom. The brush could be used to sweep out acorn mortars, baskets, and other containers. The fibers were sufficiently strong that they could be used in making hair brushes.

The Cahuilla quite possibly may have used the young spring shoots of the soap plant as a pot herb, and its saponaceous materials for a dandruff shampoo, and as a stupefying agent placed into streams to catch fish. These uses have been reported for Indian groups that were neighbors of the Cahuilla. Romero (1954:39) reported "*mo-cee*" as the Cahuilla name for the soap plant, but this name has not been confirmed by the authors.

Chrysothamnus nauseosus (Pall.) Britton (Compositae)
Ssp. albicaulis (Nutt.) Hall & Clem.

Common Name: *Rabbit Brush*   Cahuilla Name: *tesinit* [?]

Barrows (1900:79) reported that a tea made from the twigs of this shrub provided relief from chest pains and coughs. Mrs. Katherine Saubel said twigs were boiled into a tea to relieve toothache. Bright (1967:xxviii) questioned Barrows' recording of the name "tes-i-nit" for this plant, suggesting that it may be an incorrect identification based on confusion with tesinat or poppy.

Cirsium Drummondii T. & G. (Compositae)

Common Name: *Thistle*   Cahuilla Name: *N.D.*

The Cahuilla ate the bud at the base of the thistle. The method of preparation is no longer remembered.

Cirsium occidentale (Nutt.) Jeps.' (Compositae)

Common Name: *Thistle*   Cahuilla Name: *ya'i he'ash*

The Cahuilla name for this plant, a descriptive term meaning "wind-pet," has not previously been recorded. No usage is presently associated with the plant.

Citrullus vulgaris Schrad. (Cucurbitaceae)

Common Name: *Watermelon* Cahuilla Name: *istochen, estuish* [?]

Although an Old World plant, watermelons appear to have preceeded the Spanish advance into the Southwest. Father Kino (Bolton 1919:I, 249) reported watermelons being grown by the Indians at Las Sandias near the Gila junction as early as 1700. Other Spanish explorers observed watermelon being grown by Colorado River tribes, including Juan Bautista de Anza (Bolton 1930: III, 42), whose expedition was offered three thousand watermelons by the Yuma Indians. Gifford (1931:21) reported that watermelons were cultivated by the Kamia of Imperial Valley, the tribe adjacent to the Cahuilla on the east, although he supplied no evidence that went beyond the historic period.

Watermelons were mentioned as being grown by the Cahuilla near present-day Thermal by Don José Maria Estudillo, a member of the first Spanish expedition to cross the Coachella Valley (Bean and Mason 1962:46). There is a possibility that the watermelon reached the Cahuilla Indians from Indian groups to the east prior to Spanish arrival in California (see Appendix).

Harrington (unpublished) in his field notes gave *estuish* as the Cahuilla word for watermelon, which he recorded as meaning literally "something sweet." Patencio (1943:25) gave the name for watermelon as *istochen*, but since his personal reminiscences were put together by Margaret Boynton, a non-linguist, it seems possible that this was a misrendering of the plural *estuishem*. Cahuilla interviewed by the authors could not recall a name for watermelon other than the Spanish *sandia*. Mrs. Alice Lopez thought that *estuish* might refer to a squash.

The Cahuilla ate watermelons fresh or cut peel into strips and dried them for winter use. They may also have buried watermelon in sand for short-term storage as was done by some of the Colorado River tribes.

Condalia Parryi (Torr.) Weberb. (Rhamnacea)

Common Name: *Crucillo, Wild Plum*   Cahuilla Name: *chawaxat*

This spiny, intricately branched shrub grows on dry slopes and in canyons along the western edge of the Colorado Desert, usually below 3,000 feet. The shrub produces a small, red edible drupe, which was eaten fresh or dried and ground into flour for mush. The edible nutlet of the drupe was ground and leached to produce a tasty flour. Patencio (1943:88) reported that among the Palm Springs and Wanikik Cahuilla major gathering areas for wild plum were Blaisdell, Chino, and Snow Creek Canyons.

Barrows (1900:60) recorded *"o-ot"* as the Cahuilla name for wild plum. Modern Cahuilla do not recognize this name, and since Barrows (1900:36) had earlier given *"o-ot"* as the name for *Adenostoma fasciculatum,* his identification probably is in error.

Conyza canadensis (L.) Cronq. (Compositae)

Common Name: *Horseweed*   Cahuilla Name: *N.D.*

Alice Lopez recalled that this leaf annual—a native of the Old World—was boiled to make an infusion for curing diarrhea.

Croton californicus Muell.-Arg. (Euphorbiaceae)

Common Name: *Croton*   Cahuilla Name: *te'ayal*

This perennial, usually found in dry sandy areas below 4,000 feet, was used as a medicine. Mrs. Katherine Saubel was "doctored" with this plant as a child for an earache. She recalled that her mother called in a *puul* (shaman), who mashed and cooked stems and leaves of the plant into a decoction that was placed in her ear while warm.

Mrs. Saubel uses the plant for relieving congestion caused by colds. A thimbleful of the decoction is taken while very hot. The plant is toxic and used only in a small dosage.

Cucumis melo L. (Cucurbitaceae)

Common Name: *Muskmelon*   Cahuilla Name: *N.D.*

The muskmelon was among cultivated crops grown by the Cahuilla early in the post-contact period, and it may even have been grown aboriginally (see Appendix for a discussion of Cahuilla agriculture). An Old World crop, muskmelons probably reached various Indian groups of the Southwest ahead of the Spanish advance. As early as 1697, Mange (1926:256) reported that the Piman Indians harvested both muskmelons and watermelons (*melones y sandias*). Sedelmayr (1939:108,110) noted melons and watermelons being grown along the lower Colorado in 1744. In the same year, Anza recorded cultivation of both melons and watermelons by the Yuman Indians (Bolton 1930: III, 42, 51, 227-28, 323). There was thus an opportunity for muskmelons to reach the Cahuilla by diffusion from Indian groups to the east prior to establishment of the first mission in California at San Diego in 1769.

The earliest report of muskmelons being grown by the Cahuilla was made by the first Spanish expedition to cross the Coachella Valley. In 1823, Don José Maria Estudillo, a member of the Romero party, observed Cahuilla planting muskmelon (*melones*) near present-day Thermal (Bean and Mason 1962:46). In approximately the same area, the Pacific Railroad Survey expedition of 1853 under Lt. R. S. Williamson (1856:99) also noted that the Cahuilla encountered "appeared to have a good store of grain and melons, which they had raised in the vicinity."

Cucurbita foetidissima HBK. (Cucurbitaceae)

Common Name: *Calabazilla,*   Cahuilla Name: *nekhish*
            Wild Squash

The wild squash grows at altitudes between 2,000 and 4,000 feet throughout Cahuilla territory. Seeds were collected annually during the spring and ground into flour for a mush. Although not a major part of Cahuilla diet, the plant was a valuable food source, since the seeds contained about 33.8% protein and 33.9% oils.

The plant is most frequently remembered today for its saponin content, which made it useful as a soap. Jaeger (1958:283) reported, however, that the soap is so strong that particles tend to cling to clothes, causing skin irritation. All plant parts yield saponin when rubbed in water. The root and pepo (or squash) were cut into small fine pieces and used as hand and laundry soap. Clothing was placed on a smooth rock or piece of wood and plant parts were rubbed on them like laundry soap. The plant was also used as a bleach; plant parts and the material to be bleached were soaked in water for a long period. The squash and roots of the plant were gathered in quantity and stored until needed. The shell of the fruit was ground up and used as a hair shampoo.

The roots and green fruit were also used medicinally. The root was mascerated and applied to ulcers. The pulp of the squash was used as a medication for open sores. The squash was also crushed and applied to saddle sores of horses. Barrows (1900:80) observed a Cahuilla friend, Celestin Tortes, curing his horse in this manner. The dried root was boiled in water and administered as an emetic or physic.

The yellow blossoms of the plant were used as a dye. Dried gourds were used to make ladles, syringes for feminine hygiene, and very occasionally rattles. These gourds were not considered ideal rattle material, however.

Cucurbita moschata Duchesne (Cucurbitaceae)

Common Name: *Cheese Pumpkin,*   Cahuilla Name: *N.D.*
              Cushaw Pumpkin

This species designation can only be considered tentative. Nineteenth century literature on the Cahuilla refers frequently to their cultivation of both "pumpkins" and "squash," but the terms appear interchangeable and tell us nothing about the species or varieties grown. Gifford (1931:21) reported that the Kamia, the nearest neighbors of the Cahuilla to the east, cultivated *C. pepo* L. Castetter and Bell (1951:109-115) were of the opinion, however, that the bulk of pumpkins grown by Indian groups along the Colorado River and adjacent tribes were probably *C. moschata*. Their argument is well presented and worth reading.

The pumpkin may have reached the Cahuilla from neighboring tribes to the east prior to Spanish contact (see Appendix). Alarcon in 1540 observed pumpkins being grown by Indians along the lower Colorado River (Hammond and Rey 1940: 136, 139, 144). Later explorers, including Kino, Anza, and Font, reported pumpkins being grown among the Colorado River groups. A discussion of the evidence for aboriginal agriculture among the Cahuilla is presented in the Appendix.

The pumpkin was first reported as grown by the Cahuilla by José Maria Estudillo, who in 1823 noted that it was one of the crops being cultivated at a rancheria near Thermal (Bean and Mason 1962:46). The B. D. Wilson report of 1852 on Indian tribes of southern California stated that Cahuilla mountain villages had a "moderate supply of wheat, corn, melons, and pumpkins (Caughey 1952:27). Lt. Williamson (1856:98) reported "squash" as one of the plentiful crops that the Cahuilla were eager to trade with members of the Pacific Railroad Survey party in 1852.

Pumpkins were cooked and eaten fresh by the Cahuilla or cut into strips and dried in the manner of the Colorado River Indian tribes. Most names given by Cahuilla interviewed for squash were of Spanish origin, including *calabacitos*, *calabas*, and *calabaza*. Mariano Saubel recalled *paxhushlam* as a name for the crop, and Mrs. Alice Lopez responded with *estuish*.

Cuscuta californica H. & A. Cuscutaceae)

Common Name: *Dodder*   Cahuilla Name: *wikat*

Various Cahuilla gave the name *wikat* for this parasitic plant found on herbs and shrubs up to the 8,200 foot level. Mrs. Alice Lopez referred to them as *mulyak hew'*, however, which means "lizard's web." She recalled that handfuls of the plant were gathered and used as scouring pads for cleaning.

Dalea Emoryi Gray (Leguminosae)

Common Name: *Indigo Bush*   Cahuilla Name: N.D.

This densely branched shrub is found in dry, open places below 1,000 feet. Palmer (1878:654) and Saunders (1914:75) recorded its use among the Cahuilla as both a dye and medicine. Chase (1919: 67) also reported a medicinal use. Merrill (1923:236) noted that the stalk of the plant was used as a foundation element in making baskets.

Concerning the use of the plant as a dye, Palmer (1878:654) wrote: "Branches of this plant steeped in water form a light yellowish-brown dye, and emit a strong rhuelike odor. The Cahuilla Indians of California, to ornament their baskets of a yellowish-brown color, steep the brushes in a dye of that color, prepared from the Daleas."

Datura meteloides A. DC. (Solanaceae)

Common Name: *Datura,*   Cahuilla Name: *kiksawva'al*
Jimsonweed, Thorn-Apple

Datura is one of the most universally used hallucinogenic and medicinal plants known to man. Use of the plant spans man's written history, and it appears on every continent in some form. Nearly all of the tribes of southern California employed datura, including the Luiseño, Gabrielino, Chumash, and Dieguefio. In the past few years, the plant has received considerable attention from anthropologists and other scholars concerned with the various means man has employed in seeking to expand his consciousness or explore inner worlds and other realities. Such works as Weston Le Barre's *The Peyote Cult* (1959) and Carlos Castaneda's *The Teachings of Don Juan* (1968) and *A Separate Reality* (1971) have focused interest on datura and other psychotropic drugs and intrigued thousands of readers. The three books mentioned offer an excellent introduction to the ways in which hallucinogenic plants have been used and the philosophical and cosmological constructs of societies employing them.

An erect, widely branched perennial, datura may be found in sandy, gravelly open areas below 4,000 feet throughout southern California. The plant is readily recognized by its dark green leaves and whitish, bell-shaped flowers. Since all parts of the plant contain the chemical constituents responsible for datura's psychotropic effects, the Cahuilla utilized all of the different plant parts depending on the effect desired. Roots were most commonly used in a drink served at rituals; leaves were generally smoked; and both leaves and roots were crushed with other parts and mixed into a medicinal paste.

Datura is an extremely poisonous plant, and all of the Cahuilla who discussed it with us stressed that the plant is unpredictable and warned against its use by the casual experimenter. The plant may result in serious mental disorientation, disorders in locomoter activies, acute cardiac symptoms endangering heart functions, and other severe physical problems. Effects may range from temporary psychoses to death. In the past few years, several non-Indian young people in southern California have died after experimenting with the drug. Many others have required hospitalization.

The primary attraction of such a dangerous drug—and even in aboriginal times the Cahuilla well recognized such danger—was that it offered a means of coming in contact with the sacred world. Lewin (1964:94-95) in discussing the effects of datura and other psychotropic drugs speaks of them as "evoking sense-illusions in a great variety of forms, of giving rise in the human soul as if by magic to apparitions whose brilliant, seductive, perpetually changing aspects produce a rapture which is incessantly renewed and in comparison with which the perceptions of consciousness are but pale shadows."

Shamanistic Uses of Datura.—For the Cahuilla *puul* (shaman), datura offered not only a means to transcend reality and come into contact with specific guardian spirits (*nukatem*), but it also enabled him to go on magical flights to other worlds or transform himself into other life forms such as the mountain lion or eagle. Magical flights were a necessary and routine activity for Cahuilla shaman. A shaman might use the drug to visit the land of the dead, returning to the profane world with information useful to his people, or he might pursue a falling star to recapture a lost soul and return it to its owner.

Melba Bennett (n.d.) was told by a Cahuilla authority: "Medicine men use it. They were permitted to go from this world to another world. That's how they went. They would take a portion of what we call the dream weed, and then would tell people what they'd see: fantastic colors, horses, and other things. They would go clear out. Everyone was not allowed to do that."

Kroeber (1908:65) reported that it was believed that objects or events seen in datura-induced visions would come true. He added: "It was especially believed that the use of the jimson-weed would bring riches, no doubt in connection with the general idea that it conferred power and the attainment of desire." Shamans sometimes carried portions of the datura plant as talismans. The late Salvador Lopez, a Cahuilla shaman, frequently carried a vial of datura for "good luck."

Although many shamans used datura, it was not considered absolutely essential to their role and some never employed it. Even among the Cahuilla with their long experience with the drug, accidents still occurred. Kroeber (1908:66) noted: "It is said by the Cahuilla that the amount of extract of the root that is drunk must be judged by a man experienced in its use, and that a number of deaths have resulted from taking excessive quantities." Hooper (1920:347) was told by Juan Lugo of Agua Caliente reservation at Palm Springs "that several men had died as a result of drinking too much *toloache* [datura] or of eating the wrong thing afterwards."

Medicinal Uses.—Datura was employed by the Cahuilla for a variety of medicinal purposes. It was recognized as an effective pain-killer in setting bones, alleviating pain in specific areas of the body, and in relieving toothaches and swellings. For such purposes, leaves were reduced to a powder with other plant parts and an ointment was made. The ointment was applied to the afflicted area, where it relieved suffering for several days while healing took place. Often the salve was used as a hot poultice. The paste was applied to the body and heated with a hot rock held near the skin.

A medicinal paste of datura was also used to cure bites of tarantulas, snakes, spiders, and various insects. The paste was believed to possess antivenin properties against poisonous bites. More recently, datura paste was used to alleviate saddle sores on horses.

Datura leaves were steamed and the vapor inhaled by patients suffering from severe bronchial or nasal congestion. One of the chemical constituents of datura is atropine, which dilates and dries out bronchial passages and has a long history in the relief of asthmatic symptoms.

Shamans used datura in small amounts to diagnose ailments of patients. The drug permitted them to "see" the pain or disease causing difficulty and detect if witchcraft had been used against a patient. It also helped them divine cures for diseases.

Ceremonial Uses.—Datura was used ritually in puberty ceremonies held for boys to enhance their socialization as young adults and provide them with an awareness of their potential for fulfilling future functions as adults. While a young man was under the influence of datura, particular talents and supernatural connections were established which would affect him and his future role in the community.

When one considers that datura results in mental images of tremendous intensity, it is no wonder that a Cahuilla boy after his first vision under its influence became a firm believer in mythic traditions. Datura enabled him to glimpse the ultimate reality of the creation stories in the Cahuilla cosmology. The supernatural beings and aspects of the other world that he had been told about since childhood were now brought before his eyes for the ultimate test—his own empirical examination. He has seen them. They are real. This reality—which a contemporary writer (Castaneda 1968, 1971) has referred to as "non-ordinary reality"—must be perceived to be believed and accepted. Once the Cahuilla neophyte was convinced by his own perceptions, he was thenceforth locked into the entire Cahuilla cosmology, dramatically, with community guidance and support. Failure on the part of a neophyte to perceive this reality did not negate its existence, but simply meant that he was unfortunate in not receiving sufficient reality of power to understand all that the creator had made available. The degree to which one perceived non-ordinary reality was directly associated with the ability to acquire and use supernatural power within one's community.

The ritual use of datura in puberty rites for boys has been described by several anthropologists (Kroeber 1908; Hooper 1920; Strong 1929; and Bean 1972). William Duncan Strong (1929:173-175) has provided the most thorough description of datura use in the puberty ceremony. His description is presented below in its entirety:

Among the Mountain Cahuilla we encounter the jimsonweed or toloache cult in its central manifestation, to wit, the jimsonweed-drinking boys' initiation, or manet ceremony. Probably this rite occurred among the more westerly clans of the Pass Cahuilla but traces of it there are faint. All the Mountain clans, however, seem to have had it in a more or less complete form. Informants from the northern mountain clans said that manet meant "grass that could talk," but could only be heard by shamans. The Los Coyotes canyon people claimed that manet "belonged to the water," and that all manet songs were not in the iviat (i.e., Cahuilla) language but were in the "ocean language" and no one could understand them. The songs were sung to the "great witch doctors" who lived on the ocean floor, and they were prayers for the ocean winds to blow clouds over the mountains. They believed that "the ocean was above, below this were all the winds and on the bottom were the great pualem (shamans) and other monsters." The jimsonweed was a great human pul (shaman) with whom they could talk. Thus among these southern clans at least, and probably among the other Mountain Cahuilla clans, the manet ceremony was held as a prayer when water was short and food scarce, or when an epidemic raged among the people. The pualem (shamans) were always active at this ceremony. Besides this of course, it was also a boys' initiation rite.

Manet occurred every few years when the occasion demanded it, or when there were several boys to initiate. The southern clans performed manet in connection with hemwek'luwil, a three- or four-day ceremony in which small boys from six to ten years of age were taught their own clan songs and their "enemy songs" by their fathers. This took place in the wamkic [ceremonial house] or its environs, and while each boy's instruction was in the hands of relatives they were presided over by the paha [ceremonial leader]. The paha prepared strings of woven reeds called wic, and strings of eagle and flicker feathers which were worn by the dancers. A dancing leader or manet-dancer called tcauinitem was selected by the net as the best dancer to lead the boys. The southern clans called this man tcenevac. His duties consisted in leading the boys in their dances in the wamkic at night, and during the day in seeing that they practiced their songs in a secluded place away from the village. While the boys were dancing in the wamkic their relatives threw baskets and other gifts over their heads to be gathered up by the guests. This part of the ceremony was watched by clan members and visitors of both sexes.

Then came the esoteric part of the ceremony called kiksawel, "the drinking," which occurred inside the wamkic. No women or children were permitted to witness this; only the men of the clan and the novitiates, youths of eighteen to twenty years, were present. The manet-dancer whirled the bullroarer as a warning to the uninitiated to stay away. The bullroarer was called melawhic by the northern clans and meulakpic by the southern clans. The net prepared the jimsonweed, "cooking it" (meaning probably drying it), and ground it up in a small ceremonial mortar called takic, with a small pestle called paul. Water was added and the liquid was then put in a red pottery bowl called tesnut kumuismul and the paha gave each boy a swallow. The men in the wamkic then took each boy by the waist and danced around the fire led by the manet dancer. All were naked, and according to Alec Arguello old people sometimes fell into the fire in the excitement but were not burned. The novitiates became unconscious and were left in the dance house all night. The next afternoon they were taken out of the dance house and hidden in a secluded canyon by the paha. Here they were taught songs while at night for one week they danced every evening. The jimsonweed however was drunk only once.

The drinking of jimsonweed produces visions, but no especial dream cult or interpretation is remembered by modern informants. One northern Mountain Cahuilla informant said that if any boy saw in his dreams an animal that spoke to him, bad luck for his relatives would result. A southern Mountain Cahuilla informant said that boys at this time "had dreams like pualem (shamans)" and that anything seen in the vision was their "spirit or friend." Since it is fifty or sixty years since this ceremony took place among the Mountain Cahuilla it is not strange that the details are vague.

The last afternoon of the week the ground-painting or some equivalent to it was made. Alec Arguello said that among the Mountain Cahuilla clans which moved to San Bernardino the following occurred. The paha marked off a special area in front of the wamkic beyond which outsiders could not come. Then coming from the interior of the wamkic he brought a very sacred, red, white, and black basket called neat, which he carried around the space. He then returned it into the inner room of the dance house. This was very important, according to Alec Arguello, but no one knew the meaning of it. The true ground painting must have occurred there, however, for Nina Cosesos said she had seen it made three times, once at pulatana (near San Bernardino) when she was a girl of about 10 or 12 years, once at Soboba, where it was made by a Mountain Cahuilla net [the key ceremonial leader], and once at paui (Cahuilla) when she was about 16 years of age. She is a very old woman, probably between 80 and 90, hence this occurred in the neighborhood of seventy years ago. Her memories were somewhat vague, but leave no doubt as to the general nature of the ritualistic performance. The net, in the case last cited, made a shallow pit four or five feet in diameter. In this was placed a "web" of red, pauisvul, black, tul, and white, tewic, colors. These colors were made of red ocher, iron oxide, or some similar mineral, a black mineral probably graphite, and white clay. They were arranged like the spokes of a wheel within the pit. The net then explained to the boys who were being initiated the meaning of the design, but my informant did not hear this. The occurrence of the ground painting at Cahuilla is likewise remembered by Cornelio Lubo when he was very young; but he said that only certain of the old men knew what it meant, the younger being in complete ignorance. If there was any direct connection of this phenomenon with Chungichnish it was not remembered by any of the Cahuilla talked with. Primarily, manet was a boys' initiation ceremony and when the decoction had been drunk, the songs and dances learned, and the ground painting made and explained, the boys were regarded as men and full-fledged members of the clan. The ceremonial ground painting probably occurred among the southern clans, as it was well known among the Cupeno, their neighbors to the south, but no informants who remembered it among the southern Mountain Cahuilla were encountered.

Strong's reference in the last paragraph to "Chungichnish" relates to the shaman-god Chingichnich, around whom a cult sprang up among the coastal Indians and spread to some Cahuilla groups in the Santa Rosa Mountain. The use of datura was a central element in the ceremonies of this cult. The cult was keenly developed among the Luiseño and appears to have spread from them to the Cahuilla. Although most writers have argued for a Gabrieliño origin, recent studies carried out by Thomas Blackburn suggest that the cult may have originated with the Chumash.

In addition to its use in the puberty ceremonies, datura was also used by participants in various other rituals. Dancers used it to acquire great stamina and the strength necessary for some performances, particularly in the "whirling," "morahash," or eagle dance (see Strong 1929:179-180). Several Cahuilla remember that women dancers were occasionally permitted to drink a light portion of datura to facilitate dancing at rituals. Singers who were required to sing for many hours without rest often took datura to give them strength. Melba Bennett (n.d.) was told: "The dream weed is used in small amounts whenever a dance or ceremony requires an abnormal amount of strength . . . A small amount is put in the coffee. It gives you great strength, like marijuana; doesn't act like a sedative. Secret of using it: at midnight dinner drink it in coffee and dance until the sun comes up. You can feel it wear off, but when you first start, you are floating. Everybody is in about the same rhythm in dancing. You can see before midnight when people are dancing before they take it, [that] they do not dance so well together."

Other Datura Uses.—The drug's ability to enhance mental perception caused it to be used frequently in the gambling game of peon, a Cahuilla hand game played between two teams, which required great energy and acute visual perception in semi-darkness. The use of datura in this context involved not only chemical stimulation of perception, but also conferred supernatural power which was sought for winning the game.

Hunters also used datura on long treks to increase their strength, allay hunger, and acquire power that would enable them to capture game.

Descurainia pinnata (Walt.) Britton (Cruciferae)

Common Name: *Tansy-Mustard,*   Cahuilla Name: *asily*
              Pepper Grass

This annual is found throughout Cahuilla territory in dry, sandy places below 8,000 feet. It has a reddish-brown seed, which is round and flat in shape. Leaves are gathered in the spring for use as pot herbs. The ground seeds are used to flavor soups or used as a condiment with corn. The ground seeds were also used as a medicine for stomach ailments.

Barrows (1900:65) said the seeds were "ground up, cooked in a large quantity of water, and eaten with a little salt." Calistro Tortes remembered gathering the plant in summer near Rockhouse Canyon village. Hall (1902:82) reported that the Cahuilla used two other species of the genus *Descurainia* for food, but these have not been identified.

Distichlis spicata (L.) Greene var. **divaricata** Beetle (Graminea)

Common Name: *Salt Grass*   Cahuilla Name: *simut* [?]

Salt grass was not used as a food, but provided a secondary source for the condiment salt. Salt was usually obtained from the ashes of the plant after it was burnt, but sometimes salt was collected on the plant itself. The plant was cut and the salt detached by beating the plant. In general, however, the primary source of salt for the Cahuilla was natural salt found along the Salton Sea area of the Coachella Valley.

Salt grass was also used as a cleaning agent. The stiffness of the plant made it an excellent brushing material for cleaning implements or removing cactus thorns from objects. Several Cahuilla gave the name of the plant as *simut*, which is also a general name for grass. Cinciona Lubo recalled a plant common to the Salton Sea region, which she referred to as a salt plant (*samat ingily*) and said it was about three inches high. The authors were unable to obtain a specimen for identification, but it may be the same plant.

Heizer and Rappaport (1962:146-148) noted that NaCl (table salt) can be found both on the plant surface and within the plant. They also found evidence of potassium salt in some species of the genus.

Dudleya sp. (Crassulaceae)

Common Name: *Live-Forever,*   Cahuilla Name: *ya'ish*
Pygmy Weed

Various species of these perennial herbs grow in Cahuilla territory, often in rocky crevices in damp areas ranging from the Upper Sonoran to the Canadian-Hudsonian zones. Considered a delicacy by the Cahuilla, the young plants were widely sought in spring and early summer.

Each plant produces a dozen or more fleshy leaves of up to 5 inches, which are eaten raw. Though the flowering stems do not produce much food, their delicious and lingering sweet taste made them a flavored plant part also. The authors have observed species of the genus in considerable profusion from Banning to the Anza area.

Echinocactus acanthodes Lem. (Cactaceae)

Common Name: *Barrel Cactus*   Cahuilla Name: *kupash*

This cactus provided both food and water when needed for the Cahuilla. Found on gravelly fans and rocky slopes of the Colorado Desert upwards to 5,000 feet, barrel cactus are particularly abundant at the openings of canyons, often near traditional village sites. They are heavily distributed in the San Gorgonio Pass near Chino Canyon and on the western and southern slopes of the Santa Rosa Mountains.

The buds of the plant were a Cahuilla food source. In the spring, a dozen or more budding flowers grow in a circle about the top of the plant. Ripening time for the buds varies from the lower to higher desert areas ;thus the buds provided several months of fresh food within the territory of most Cahuilla groups. The *Wanikik* Cahuilla of the San Gorgonio Pass, for example, gathered these buds in late April near Whitewater Canyon and in late July near Banning and Beaumont. The Palm Springs Cahuilla and other desert groups probably had a shorter gathering period, beginning in April on the desert and extending perhaps as late as June in the low foothills of the Santa Rosas. The different collecting areas were visited repeatedly until the crop was exhausted.

The buds were gathered by women, each one plucked from the cactus with a pair of short sticks and then placed in a gathering

67

basket. The sharp spines of the cactus made the use of the gathering sticks necessary to protect the collector from injury.

The buds can be eaten fresh, but tend to have a bitter taste. To release this bitterness, the buds were usually parboiled several times (after removal of the base of the bud). After parboiling, the buds were eaten or dried in the sun for storage. When stored buds were needed, they were simply recooked in water. Salt was usually added to the water to enhance the taste.

Melba Bennett (n.d.) recorded another method of bud preparation. A pit was dug and a mesquite fire was started in the bottom of the pit. When the fire was reduced to coals, rocks were added and covered with slightly damp sand and leaves (or cloth in recent years). The fruit was then placed on the layer of leaves, covered with another leaf layer, and then sand was placed over the leaves. Another fire was built on this top layer of sand. The buds were allowed to steam cook for several hours. The taste of this cooked fruit was compared to artichoke. The buds were also used in stew or eaten with mountain sheep or jackrabbit meat. The steamed fruit could also be dried in the sun until dehydrated and stored through several seasons.

The mature flower of the plant was also eaten. It was cooked, prepared, and preserved in the same manner as the bud. It had a more bitter taste, however, even when parboiled, not unlike that of brussels sprouts.

More recently, a favorite preparation method has been to fry the bud, sometimes mixing it with chile and other condiments. A generation ago, Cahuilla people also began using modern canning methods to preserve the fruit. Mrs. Katherine Saubel recalled that her family canned over one hundred quarts of buds after one spring harvest.

The barrel cactus also provided a desert reservoir, one which has long been familiar to many desert travelers at times of emergency. To obtain water, the top of the cactus is sliced off, a portion of the pulp is removed to create a depression, and then the pulp is squeezed by hand in the depression until water is released from the spongy mass. Romero (1954:47) said the flesh of the fruit could also be used as a thirst quencher.

The body of the plant was of sufficient strength and size that it could be used as a cooking vessel. The top was cut off for this purpose and the interior was dug out. Water was then put into the depression and heated with hot stones.

Echinocactus polycephalus Engelm. & Bigel. (Cactaceae)

Common Name: *None in Munz*   Cahuilla Name: *u'ush*

This cactus, found on rocky slopes between 2,000 to 5,000 feet, provided an edible bud (Barrows 1900:68). It was harvested between May and June, depending upon climactic factors that influenced its ripening in different areas.

Elymus condensatus Presl. (Gramineae)

Common Name: *Rye Grass, Cane*   Cahuilla Name: *pahankish*

This perennial rye grass species is common below 5,000 feet throughout much of southern California. The stalks served two important functions: arrowshafts and roof thatching.

In arrowmaking, rye grass stems were fire hardened to serve as the main shaft. A foreshaft of some other material was then inserted at one end and secured with asphaltum or pine pitch. The Luiseño, who also used rye grass in arrowmaking, employed *Adenostoma fasciculatum* for the foreshaft. Occasionally, painted arrows of rye grass were used in ceremonial dances. The only known example of such an arrow has been exhibited at Palm Springs Desert Museum. Possibly some specimens have survived in the hands of private collectors.

Although the northeastern neighbors of the Cahuilla, the Southern Paiute, used the plant as a source of food, no Cahuilla questioned by the authors could recall such usage.

Encelia farinosa Gray (Compositae)

Common Name: *Brittle-Bush, Incienso*   Cahuilla Name: *pa'akal*

This low, branching shrub is found on dry stony slopes up to 3,000 feet throughout Cahuilla territory, particularly on the desert. Chase (1919:82, 99) reported that the Cahuilla used a gum of this plant as a medicine. The gum was heated and applied to the chest to relieve pain. Mrs. Katherine Saubel recalled that her grandmother boiled the blossoms, leaves, and stems to make a decoction that was held in the mouth to relieve toothache.

Ephedra nevadensis Wats. (Ephedraceae)

Common Name: *Mormon Tea,*   Cahuilla Name: *tutut*
Miners' Tea, Mexican Tea

This gymnosperm, distantly related to the pines, was particularly popular among the Cahuilla and other southwestern Indian groups for its use in a refreshing beverage. *E. nevadensis* and other southwestern species such as *E. californica* Wats., and *E. viridis* Cov., were later used for the same purpose by emigrants, hence the common names given above.

Barrows (1900:73) reported that ephedra grew along mountains facing the desert and was especially abundant on the slopes below Piñon Flats. The plant was so popular that its twigs from which the tea was made could "almost universally be found tucked away among the thatching of every jacal or packed away in basket and olla." The seeds were also ground into a meal and eaten as a mush.

The tea, which was considered to be both a beverage and a medicine, was prepared by boiling fresh or dried twigs in water until a "wine-colored" brew was achieved. The plant was picked in late summer when mature, usually by women, although men also collected it. The twigs were stored indefinitely.

Hardin and Arena (1969:130) warn that the plant should be used with caution and list it among their toxic plants.

Chase (1919:75) reported that it was used by southern California Indians for stomach and kidney ailments. Palmer (1878:653) gave it the name of *Ephedra antisyphilitica* or teamster's tea and said it was a common Indian remedy in the southwest against syphilis and gonnorhea. Romero (1954:22) reported that an infusion prepared from leaves and blossoms could be used for flushing the kidneys.

Cahuilla interviewed by the authors were unfamiliar with the plant as a remedy for venearal diseases. Among Cahuilla it is generally believed that the tea purifies the blood and "clears the system." It was agreed that if the tea is taken for too long it is "bad for the system."

Equisetum L. (Equisetaceae)

Common Name: *Horsetail, Scouring-Rush*   Cahuilla Name: *N.D.*

Various species of this genus, which is a spore-bearing plant like the fern, were used by Indians for curing ailments such as kidney stones and dysentery. An astringent, it has been said to be useful in cleaning external ulcers and facilitating urinary secretions (Martinez 1944:293).

Alice Lopez recalled that an unidentified species of *Equisetum* was used by the Cahuilla in the manner of a cleaning pad as a cleansing agent. The medical uses were not remembered by present-day Cahuilla. Curtis (1926:121) recorded a Cahuilla tale in which a culture hero ate some rushes (identified as *Equisetum*), vomited them, and became well. The tale suggests that at one time the Cahuilla may have used this plant as an emetic.

Eriodictyon trichocalyx Heller (Hydrophyllaceae)

Common Name: *Yerba Santa*   Cahuilla Name: *tanwivel*

Yerba santa, an important medicinal plant among the Cahuilla, may be found up to 8,000 feet on dry rocky slopes and fans, often in the chaparral or pinyon-juniper regions. Palmer (1878:651) first noted its use as a "great medicine" among the Indians of California as an internal and external remedy for rheumatism and for partial paralysis. He also reported that for lung problems the leaves were chewed dry, smoked, or made into a tea. Barrows (1900:78) reported that among the Cahuilla the leaves were "pounded up and are bound upon the sores of both men and beasts and a strong decoction is used for bathing sore parts or the limbs when painful or fatigued."

The Cahuilla today use yerba santa for the same purposes recorded by Barrows and Palmer. The thick, sticky leaves, either fresh or dried, are boiled, mixed with sweetening agents such as honey to eliminate bitterness, and drunk in a tea. The decoction made by boiling the leaves may take the place of regular beverages or serve as a medicine. As a medicine, it is said to be a blood purifier and a cure for coughs, colds, sore throats, asthma, catarrh, tuberculosis, and rheumatism. The decoction may also be applied as a liniment to reduce fever. Fresh leaves are also pounded and placed on sore or fatigued limbs or heated and placed on the body to cure rheumatism. Leaves are sometimes chewed fresh as a thirst quencher. The most efficacious plants are said to grow in mountainous areas.

Mrs. Katherine Saubel has provided the following recipe for a cough medicine: take three leaves of yerba santa, wash well, boil, put in one-half teaspoon of sugar, and administer one teaspoon every four hours for cough.

Eriogonum Michx. (Polygonaceae)

Common Name: *Buckwheat*   Cahuilla Name: *hulaqal*

Various species of buckwheat may be found in Cahuilla territory in most of the life zones. It is especially abundant on foothill slopes, in dry sandy desert areas, and on alpine meadows. In the desert, the plant is seasonally limited to moist periods of the spring and winter. Edible shoots were available in the desert from February to May, while the seeds, also eaten, were gathered by the Cahuilla from June until September.

A strong, black decoction made from the leaves was drunk as a cure for headaches and stomach disorders (Barrows 1900:78). The white flowers were steeped to make an eye wash or a drink that was said to clean out the intestines. Leaves, growing near the root, were used as a physic. A tea made from the plant was also said to cause the uterus to shrink, inhibiting dysmenorrhea. The oldest plants are said to be most efficacious as medicine.

Eriophyllum confertiflorum (DC.) Gray (Compositae)

Common Name: *None in Munz*   Cahuilla Name: *N.D.*

Seeds of this plant, which abounds from sea level to 8,000 feet along the western edges of the Colorado Desert, were gathered by the Cahuilla as a food. Available in different areas from June to November, the seeds were parched and ground into flour. The plant is also said to have been used for medicinal purposes that are today forgotten.

Erodium L'Her. (Geraniaceae)

Common Name: *Storksbill, Filaree,*   Cahuilla Name: *pakhanat*
                    Clocks

Various species of *Erodium* may be found throughout the foothills and lower passes of Cahuilla territory as high as 6,000 feet. Particularly common are *E. texanum* Gray and *E. cicutarium* (L.) L'Hér. One of the earliest fall plants, often appearing after the first fall rains, filaree was a favorite pot herb.

Calistro Tortes said that the best time for gathering was just prior to blossom between January and April. He compared the taste of the plant to spinach. The plant was usually cooked while fresh, although it could be eaten uncooked or preserved for a brief period. The plant grew abundantly around village sites, where it was readily available to women on their daily food round. Mrs. Alice Lopez applied a different Cahuilla name to the plant, calling it *tuchily hemu'* or "humming bird's nose."

Eschscholzia Cham. (Papaveracea)

Common Name: *California Poppy*   Cahuilla Name: *tesinat*

Several species of the California poppy grow in Cahuilla territory. This annual or perennial herb may be found up to 5,000 feet in many plant communities. Cahuilla women used the pollen as a facial cosmetic. The plant was also said to provide a sedative for babies.

Eucalyptus L'Hér. (Myrtaceae)

Common Name: *Eucalyptus,*   Cahuilla Name: *qahich'a waavu'it*
               Gum Tree

This Australian native, introduced to California in the nineteenth century, was given a Cahuilla name meaning "tall thing." In recent times, the leaves were used in steam treatments to cure colds. The leaves were boiled in water and the patient held his head over the bowl. A blanket was then placed over the patient, who inhaled the steam to relieve sinus congestion.

Euphorbia L. (Euphorbiaceae)

Common Name: *Spurge*   Cahuilla Name: *temal hepi'*

Various species of spurge, both introduced and native, may be found throughout Cahuilla territory. Both native and introduced species were known as *temal hepi'* or "earth's milk." Hardin and Arena (1969:17) warn that some species of spurge may be poisonous to allergic persons, resulting in inflammation of the skin or large blisters on contact. Children are especially susceptible.

According to Mrs. Alice Lopez, both native and introduced species were used as a medicine for reducing fever and as a cure for chicken pox and smallpox. The plant was boiled and the afflicted person was bathed with the decoction. The infusion was also drunk to cure sores in the mouth.

Alejo Patencio said that *E. polycarpa* Benth. was used to cure rattlesnake bites, except for those "whom *amanah* [the Great Spirit] had called" (Strong 1929:119). Lucille Hooper (1920:352) reported that in curing snakebites the sucking method was sometimes foregone in favor of *Euphorbia* treatment. Cahuilla in the desert also told Hooper (1920:352) that "a snake before fighting a rattlesnake always eats some of this weed [*Euphorbia*] so as to be immune to the poison." When treating snakebites, a paste was made from the plant and applied as a poultice. Occasionally, a medicinal tea was made from spurge.

Mrs. Alice Lopez said that the sap of one species of spurge was used to cure earaches, provided a remedy for bee stings, and could be placed on sores to facilitate healing. Oscar Clarke identified a specimen of this plant as *E. melanadenia* Torr.

Fourquieria splendens Engelm. (Fourquieriaceae)

Common Name: *Ocotillo, Candlewood*   Cahuilla Name: *utush*

Ocotillo, a starkly beautiful desert plant, commonly occurs on the western edges of the Colorado Desert in dry, rocky places up to about 2,500 feet. It provided both food and fencing material for the Cahuilla.

From March until the middle of summer, the ocotillo provided edible blossoms, forty or fifty to a plant. The blossoms were eaten fresh or soaked in water to make a summer drink. The drink was slightly bitter, yet pungently pleasant. Soon after blossoming, seed pods became available for more food. The seeds were parched and ground into a flour from which mush or cakes were made. Earle and Jones (1962:241) reported that the seeds have a 28.8% protein content and an 18.6% oil content.

Chase (1914:84) observed that among the Mexicans the ocotillo provided material for building fences and the stalks were used as torches. Juan Siva said the Cahuilla built similar fences around gardens to prevent rodents from attacking cultivated crops. Plant materials were utilized for firewood.

Fragaria californica Cham. & Schlecht. (Rosaceae)

Common Name: *Wild Strawberry*   Cahuilla Name: *piklyam*

The fruit of the wild strawberry has always been and still is a choice delicacy among the Cahuilla. Common to the southern California mountains in shaded, damp areas, the plant bears fruit that is available for gathering from April through June. The plant has been pointed out to the authors in the Anza area and near Warner's Ranch. The fruit was always eaten fresh.

Grindelia squarrosa (Pursh) Dunal. (Compositae)

Common Name: *Gum Plant*   Cahuilla Name: *N.D.*

Palmer (1878:652), who worked closely with the Cahuilla, reported that a decoction made from this plant was used internally by the Indians of southern California to cure colds. Since knowledge of this remedy was also widespread among Mexican-Californians of the nineteenth century, it appears probable that the Cahuilla were familiar with this usage.

Gutierrezia microcephalia (DC.) Gray (Compositae)

Common Name: *Matchweed, Snakeweed*   Cahuilla Name: *N.D.*

Mrs. Alice Lopez identified this plant as useful in the cure of toothaches. A solution made from the plant was used as a gargle or part of the plant was placed inside of the mouth periodically to ease pain.

Haplopappus Cass. (Compositae)

Common Name: *None in Munz*   Cahuilla Name: *N.D. for genus*

Two species of this shrub were used by the Cahuilla for medicinal purposes: *H. acradenius* (Greene) Blake and *H. palmeri* Gray. The former had the Cahuilla name of *mechawal* and the latter was named *henily.*

James (1906:247) reported that *H. acradenius* or the desert isocoma was one of the "chief plants used for medicine" by the Cahuilla for curing colds and sore throats. For colds, the root of the plant was cut into small pieces, which were boiled in water. The patient drank the infusion at hourly intervals. For sore throats, leaves were soaked in a pan of boiling water. The patient knelt over the pan and a red-hot rock was dropped into the liquid. A blanket was thrown over the patient's head, and he then inhaled the steam until symptoms disappeared. According to Mrs. Alice Lopez, the patient sometimes had to uncover the blanket several times to wipe away perspiration.

Chase (1919:75) also reported that boiled leaves of the same species were used as a poultice. Calistro Tortes told the authors that the poultice was used for sores, and confirmed the usage reported by James. Cinciona Lubo said that the plant was occasionally used to build fences as a protection from cold winds.

Barrows (1900:78) reported that the leaves and twigs of *H. palmeri* provided an important medicine. He said that leaves were bound upon the feet and hot stones were pressed against them to relieve swelling and pain. He also said the leaves could be used similarly to relieve swellings on the arms or reduce the possibility of infections from cuts. Occasionally, horses were washed with an infusion of *H. palmeri* to keep away insects. The best time for gathering the plant is said to be early summer.

Hardin and Arena (1969:126) warn that at least one species of *Haplopappus* is known to contain poisonous constituents that can result in liver damage and anuria.

Helianthus annus L. (Compositae)

Common Name: *Sunflower*   Cahuilla Name: *pa'akal*

The sunflower, a popular seed food throughout the United States in aboriginal times, was gathered in quantity each fall by the Cahuilla. Although some Indian tribes of the Southwest cultivated the plant, cultivation of sunflower has not been reported among the Cahuilla except in modern times.

Sunflower seeds were dried by the Cahuilla and then ground and mixed with flour from other seeds. Sunflower seeds compare favorably with olive oil in food value. Seeds analyzed by Earle and Jones (1962:247) contained 3.1 to 5.6% ash content, 26.2 to 43.8% protein content, and 28.2 to 54.5% oil content. Starch and tannin were absent, although there were traces of alkaloids.

Heliotropium curassavicum L. var. oculatum Heller (Boraginaceae)

Common Name: *Chinese Pusly,*   Cahuilla Name: *hoskos* [?]
            Wild Heliotrope

Mrs. Alice Lopez readily identified this plant as *hoskos* or *hoshos*, but she could not recall a usage. The plant was commonly found near Torres-Martinez Indian Reservation.

Hemizonia fasciculata (DC.) T. & G. (Compositae)

Common Name: *Tarweed*   Cahuilla Name: *N.D.*

Tarweed grows on arid slopes and in sandy washes and mesas of the lower and upper Sonoran life zones. The entire plant, including the seeds, was used as a famine plant by the Cahuilla. Palmer (1878:605) first recorded its usage as follows: "This plant in case of hunger is eaten by the Indians of southern California after being cooked in the following manner: A quantity of the plants are boiled down until the liquid is of a thick tarry consistency, when it is ready for the stomach of the Indians. Its tar-like taste is objected to by some."

Hesperocallis undulata Gray (Liliaceae)

Common Name: *Desert-Lily*   Cahuilla Name: *N.D.*

The tunicated bulb of the desert-lily with its garlic-like flavor was highly regarded as a food by the Cahuilla and many other Indian groups of the southwest. The plant is common to dry sandy flats and dunes on the desert, usually below 2,500 feet.

Desert-lily was usually ready to eat in early spring from February to May, although it was not a dependable plant and often was available only in wet years. The bulbs were eaten raw or baked. When baked, the bulbs were placed in a stone-lined pit, covered with hot ashes and leaves, and left to bake for 12 to 24 hours.

Heteromeles arbutifolia M. Roem. (Rosaceae)

Common Name: *California Holly, Toyon,* Cahuilla Name: *ashwet*
              Christmas-Berry

California holly with its conspicuous scarlet berry grows frequently on semidry brushy slopes and in canyons below 4,000 feet. The bush produces berries abundantly between September and February, bearing most heavily during November and December.

Hicks (1963:111) pointed out that the holly is seldom mentioned in ethnographic literature, which suggests it was not highly regarded as a food source by Indians of the Southwest. Balls (1965:36-37) reported that California Indians often cooked the berries by roasting them in the fire or tossing them on hot coals, which removed the slightly bitter taste. The Cahuilla ate the berries both cooked and raw.

Holodiscus microphyllus Rybd. (Rosaceae)

Common Name: *None in Munz*   Cahuilla Name: *tetnat*

Barrows (1900:61) described a small grayish-green shrub, tentatively identified by Jepson as *H. discolor* Maxim, which produced an edible fruit. This plant was probably *H. microphyllus*, which is found in Cahuilla territory. Present-day Cahuilla questioned were unfamiliar with its food use. Since the fruit is not fleshy, any nutrition would have been confined to the seed.

Hordeum Stebbinsi Covas (Gramineae)

Common Name: *Barley*   Cahuilla Name: *pa'ish heqwas*

This European native, introduced into the Southwest at an early date, was called *pa'ish heqwas*, meaning "field mouse's tail." The seeds were eaten occasionally by the Cahuilla when other foods were scarce.

Hordeum vulgare L. (Gramineae)

Common Name: *Common Barley*   Cahuilla Name: *N.D.*

Barley has been grown by the Cahuilla since at least the mid-nineteenth century. This European crop plant was observed being grown by the Yuman Indians of the Colorado River in 1775 by Anza (Bolton 1930: II, 177), but whether it reached the Cahuilla from Indian groups to the east or Spanish missions on the coast is unknown (see Appendix).

Lt. R. S. Williamson (1856:46) first reported barley being grown among the Cahuilla in 1853 when he crossed the Coachella Valley with the Pacific Railroad Survey. About twelve miles south of Agua Caliente (present-day Palm Springs), Williamson noted the remains of an Indian hut and the "stubble of a barley field." Further east, near Thermal, Williamson (1856:99) encountered Indians who had a "good store of grain and melons, which they had raised in the vicinity." The type of grain being grown near Thermal was not specified.

Hydrocotyle L. (Umbelliferae)

Common Name: *Marsh Pennywort*   Cahuilla Name: *N.D.*

Mrs. Alice Lopez described to the authors and to Oscar Clarke a plant used as a green, which spread thickly on top of water in various wet places. Mr. Clarke was of the opinion that this plant might possibly have been one of several species of marsh pennywort which can be found in areas of permanent water in southern California.

Hyptis Emoryi Torr. (Labiatae)

Common Name: *Desert Lavender*   Cahuilla Name: *N.D.*

This aromatic, erect shrub was used by the stepfather of Mrs. Alice Lopez for stopping hemorrhages. A prominent Cahuilla *puul* (shaman), Mrs. Lopez's stepfather boiled blossoms and leaves of the plant and gave them to patients as an infusion. By bottling the remedy, he was able to preserve it for several years.

The seeds of the desert lavender are edible, but it is not known whether the Cahuilla used them as a food.

Isomeris arborea Nutt. (Capparidaceae)

Common Name: *Bladderpod*   Cahuilla Name: *N.D.*

A common plant throughout Cahuilla territory, the bladderpod usually grows in desert washes and subalkaline places below 4,000 feet. The pods of the plant were a source of food. Barrows (1900: 66) wrote about its use as follows: "One of the most curious ways of preparing foods is the treatment of the small plump capsules of the bladder pod (*Isomeris arborea* Nutt.), a shrub with a hard yellow wood. These pods, so the Indians have informed me, are gathered and cooked in a small hole in the ground with hot stones."

Curiously, the bladderpod has been entirely forgotten by modern Cahuilla as a food source. We were unable to locate a single Cahuilla who recalled such a use. Possibly, the slight food quantity available per plant and the inconvenience of preparation may have led to abandonment of the plant's use. However, these plants do grow in great abundance; the authors have observed pods at the proper stage of growth for eating from March until May throughout the San Gorgonio Pass in the desert and foothill areas. Since various cacti and many greens are available at this time of year, it seems possible that in more recent times with increased cultivation of agricultural crops bladderpod may have been ignored for more accessible foods. The pungent smell of the plant, reminiscent of skunk odor may even have repelled people who were no longer compelled to use the plant as a food.

Juncus L. (Juncaceae)

Common Name: *Rush, Wire-Grass*   Cahuilla Name: *seily*

Various species of rush were employed by the Cahuilla in making baskets. Among these species were *J. textilis* Buch., *J. Lesueurii* Bol., *J. effusus* L. var. *pacificus* Fern. & Wieg., and *J. acutus* var. *sphaerocarpus* Engelm. The latter two species are not given varietal identifications by Merrill (1923:237), but since these varieties are found in Cahuilla territory it seems likely the fuller identification is correct (Munz 1965:1404, 1405). The *J. Lesenerii* Bol. (sic) reported by Barrows (1900:42) would appear to be *J. Lesueurii* Bol.

Rush plants were a source of several colors and provided gradations in shades that could be used to produce the elaborate designs characteristic of Cahuilla basketry. The rush was used in making open-work baskets used for collecting foods, roughly made baskets such as were sometimes used to leach acorn meal, and more finely woven baskets.

Describing the use of *J. Lesueurii*, which grew abundantly in cienegas and damp soil, Barrows (1900:42) wrote as follows: "The scape and leaves are two to four feet high, or more, stout and pungent. A supply of these tough scapes is gathered by the basket-maker and cut off at a suitable length. She then takes a rush by one end and with her teeth splits it into three equal portions, carefully separated the entire length of the piece. Each scape thus furnishes three withes. This reed is, near its base, of a deep red, lightening in color upwards, passing through several shades of light brown, and ending at the top in a brownish yellow. Thus this bulrush, in its natural state, furnishes a variety of colors." According to Barrows (1900:43), the scapes of the rush were called *se-il* and their red portions *i-i-ul*. When dyed black, the materials were referred to as *se-il-tul-iksh* or black *se-il*.

One term in the litreature that has sometimes been considered synonymous with *Juncus* is *maiswat* (Curtis 1926:21-39). The *maiswat* plant was used to make the ceremonial bundle, also called *maiswat*. If rush was ever used for this purpose, the usage was probably recent, since the Cahuilla creation myth suggests *maiswat* was a type of seaweed.

In the Cahuilla cosmogeny, *maiswat* was first utilized for making an effigy of the creator-god Mukat for ceremonies that followed his death. According to Lucille Hooper (1920:327): "When they were ready to hold the fiesta [in memory of Mukat], Coyote told them he knew what to make effigies of, and offered to go to the end of the world to get it. Misvut (a seaweed) was what he got. It grew far under the water. It had probably been made in the beginning for this purpose." Strong (1929:143) also identified *maiswat* as a seaweed, used to make seaweed matting for the "wrapping of the clan fetish bundles." According to Strong, Coyote went to the ocean, where he gathered three plants, "paña maiswut, paña hekwa, and paña wiava." The seaweed plant or *maiswat*, Strong reported, was cut and wrapped with the other two plants, but he was unable to ascertain what kinds of plants the latter two were. Strong also noted that in recent times the *maiswat* bundle usually was made from a tule or reed matting. Several Cahuilla interviewed by the authors have given their opinion that *maiswat* was different than rush, being a reedlike plant smaller than a tule.

Juniperus californica Carr. (Cupressaceae)

Common Name: *California Juniper*   Cahuilla Name: *yuyily*

California juniper is usually found in the pinyon-juniper belts of the interior upland areas of Cahuilla territory. It often grows in great abundance, particularly at about the 3,000 foot level on the eastern slopes of the Santa Rosa Mountains. Piñon Flats is a typical high concentration juniper area.

An arborescent shrub, the California juniper produces a berry which provided an irregular source of food. According to Barrows (1900:63), the berry was called *iswat*. Cahuilla interviewed by the authors gave the name *yuyily* for the plant. The berries were consumed in great quantities between June and August. Although berries were available every year, individual trees bear fruit only every other year. The trees vary in the quality of berry produced, with some trees bearing a bitter fruit.

When gathered the berries were eaten fresh or dried in the sun and preserved for future use. Dried berries were ground into a flour and made into mush or bread. Although not a major food of the

Cahuilla, the berries rounded out their yearly diet. Since juniper berries were found in the same areas as pinyon trees, whose nuts ripened about the same time, harvesting of both plants could be conveniently carried out at the same time.

Barrows (1900:33) appears to have confused the California juniper with another species, *J. occidentalis* Hook, or the western juniper. This plant grows in parts of Cahuilla territory, particularly the San Bernardino Mountains, and probably was used as a food source by the Cahuilla. Palmer (1878:594) reported that many southern California Indian groups gathered berries from this tree, but noted that the berries "have more of a juniper taste than the other species." Zigmond (1941:49) commented that the food value of juniper berries was insignificant.

Romero (1954:9) referred to the California juniper as *gla-wat-pool*, which may have been a name attached to the plant as a medicinal remedy. The berries were used for making tea or simply chewed as a cure for fevers and colds. Among other Indian groups of the southwest, the bark was also used in preparing medicine for the treatment of colds, fever, and constipation. Other tribes are also known to have used the bark for making clothing and mattresses. Use of the bark is not recalled by present-day Cahuilla.

Lagenaria siceraria (Cucurbitaceae)

Common Name: *Bottle Gourd*   Cahuilla Name: *N.D.*

Kroeber (1908:62) concluded that a cultivated-gourd rattle (species unknown) obtained from the Cahuilla was a trade item from the Colorado River, since he considered the Cahuilla to have been non-agricultural. Lawton and Bean (1968), however, have argued a circumstantial case for marginal agriculture among the Cahuilla in the aboriginal period (see Appendix). If their hypothesis is correct, then gourds might have been one of the crop plants grown.

Recently, Wilke and Fain (1972) reported their discovery of a bottle gourd in a small rockshelter several miles northwest of the site of Toro Village in the Coachella Valley. The species identification was made by Dr. Thomas W. Whitaker of the U.S. Department of Agriculture, Horticultural Field Station at La Jolla. The age of the specimen remains unknown.

The gourd when found was nearly intact, and was assumed to have been intentionally hidden in the rockshelter, since it was found in a crevice, partly concealed by several rocks. The peduncle (stem) and interior contents of the gourd were missing. Castetter and Bell (1951:116) report that such treatment was employed by the Yuman tribes to remove the contents of gourds in preparation for use as containers and rattles. Wilke and Fain (1972:2) concluded that the gourd probably served as a rattle. For the present, the significance of the find remains difficult to assess, they noted, since the whole question of aboriginal agriculture in the Coachella Valley must await the results of future archaelogical research.

Larrea divaricata Cav. (Zygophyllaceae)

Common Name: *Creosote Bush*   Cahuilla Name: *atukul*

Creosote is a dominant shrub over large areas of the desert and on dry slopes and plains up to 5,000 feet. It was one of the plants most commonly used for medicinal purposes by the Cahuilla and other southwestern Indian groups. The shrub also provided a source of fine firewood.

Krochmal, Paur, and Duisberg (1954:4) reported creosote has been used by Indians for treatment of almost as many diseases as penicillin. In 1942, they noted, scientists at the University of Minnesota first extracted a material known as "nordihydroguaiaretic acid" or N.D.G.A. from the plant, which proved to be a remarkable antioxidant (rancidity retarder) for fats and oils. Leaves of the plant contain about 16% protein, comparable to alfalfa, and it has been suggested they might make excellent livestock feed. Munz (1965:158) stated that creosote has antiseptic properties.

The Cahuilla made a medicinal tea from the stems and leaves of the creosote. The tea was used to cure a variety of ailments, including colds, chest infections, bowel complaints, and stomach cramps associated with delayed menstruation. The tea was considered a good decongestant for clearing lungs and an effective cure for cancer. The tea was often mixed with a sweetening agent and used as a general health tonic before breakfast. Honey is usually the sweetener used by present-day Cahuilla. Heavy doses of tea were used for inducing vomiting.

The leaves of the plant were boiled or heated and the steam inhaled to relieve congestion. In the early days, this was done in the sweathouse. Modern Cahuilla boil the leaves in a bowl or pot, cover their head with a blanket, and inhale the steam.

Solutions or poultices were made from creosote to heal open wounds, draw out poisons, and prevent infections. Mrs. Katherine

Saubel has used a decoction from the leaves to prevent infection in a puncture wound caused by a rusty nail. The decoction was pressed over the wound, and then heated with a hot stone. Powder from crushed leaves were also applied to sores and wounds. A liniment made from creosote was a favorite cure of elderly people afflicted with swollen limbs caused by poor blood circulation.

Romero (1954:37) reported that an infusion made from the plant was used to eradicate dandruff and was useful as a hairwash. The treatment for dandruff was carried out once a week for about two months. He also observed the use of the infusion as a disinfectant and a body deodorizer. Barrows (1900:79) recorded the use of creosote as a remedy for consumption, and also said it was given to horses with colds, distemper, or running at the nose.

Lasthenia glabrata Lindl. (Compositae)

Common Name: *Gold Fields*   Cahuilla Name: *aklukal*

This succulent winter annual grows in fields and on hillsides up to 4,500 feet, ranging from sinks to grasslands and oak woodlands. The dark, elongated seeds were gathered as food by women from May through July. The seeds were parched and ground into flour or eaten dry or mixed with water as a mush.

Lathyrus laetiflorus Greene (Leguminosae)

Common Name: *Wild Pea*   Cahuilla Name: *chawak se'ish* [?]

Cahuilla have said that this plant was used as a food, although the plant parts used were not remembered. Hardin and Arena (1969) warn, however, that the seeds of *Lathyrus* species are poisonous and can cause paralysis.

Mrs. Alice Lopez referred to the plant as *chawak se'ish* or "crawling flower," but said that there was another Cahuilla word for the plant which she could not recall. She also furnished the name *chiichara*, which is derived from the Spanish *chicharo* or "pea."

Layia glandulosa (Hook.) H. & A. (Compositae)

Common Name: *White Tidy Tips*   Cahuilla Name: *N.D.*

This vernal annual usually is found in grassy flats on the desert below 3,500 feet, but may occur up to 7,800 feet. Seeds of the plant were gathered by women from June to August, ground into flour, and cooked with other ground seeds in a mush. Since *L. platyglossa* (F. & M.) Gray is much similar and was also found in parts of Cahuilla territory, it seems likely that it also may have been used as a food.

Lepidium nitidum Nutt. (Cruciferae)

Common Name: *Peppergrass*   Cahuilla Name: *pakil*

This small erect annual was common to open places below 3,000 feet. The leaves were boiled and allowed to set until a brownish-colored solution was present. The solution was used to wash hair, keep the scalp clean, and prevent baldness. The remedy was stored in a pot and kept for use as needed.

Romero (1954:66) reported that *L. epitalum*, another species of peppergrass, was used as a reducing tea, but no such species is listed by Munz (1965). His name for this plant, *chesa-mok-ka-mok*, was not recognized by Cahuilla interviewed by the authors.

Libocedrus decurrens Torr. (Cupressaceae)

Common Name: *Incense Cedar*   Cahuilla Name: *yulil*

The incense cedar was used by the Cahuilla for constructing conical-shaped bark houses in the mountains for temporary use while they camped to gather and process acorns each October and November. In some parts of Cahuilla territory, cedar slabs were also used in more permanent construction. Barrows (1900:38) wrote as follows about such structures: "At Santa Rosa, the houses are made of slabs of cedar from the trees among which the village lies. These slabs are planted upright in the ground to form the sides, the chinks are filled with mud, and the flat or slightly pitched roof is made of poles covered with cedar bark.

Harry C. James of Lake Fulmore in a letter to the authors has provided us with an account of the construction of such a bark house by the late Calistro Tortes.

While I was talking with the late Calistro Tortes of the Santa Rosa Cahuilla one day, he mentioned the fact that his people and those of the Mountain Cahuilla sometimes built conical houses with the bark of the incense cedar.

Later, we showed him a picture of a shelter house made by the Indians of Yosemite Valley, which he stated was a bit small and more steep-walled than those made by the Cahuilla. The photograph of one such structure, which had been found some years ago on Mt. Palomar by the late Guy Fleming was, he thought, very similar to those of the Cahuilla.

Calistro agreed that if the boys then attending our Trailfinders Camp and I would assemble enough bark he would be glad to come over to our place in the San Jacinto Mountains and help us build such a bark "kish."

He had stated that the Cahuilla usually did not try to strip the bark from living trees, as this was a difficult task. By searching around through the mountains, they could usually find all the slabs of bark they needed, which were easy to remove from dead trees. As the incense cedar bark resists rot, it remains in good condition for many years after the death of the tree.

We collected slabs of bark of all sizes, from three or four feet in length and about six inches wide to large pieces a foot or two wide and from twelve to fourteen feet long.

The diameter of the house that was built with Calistro's help was fifteen feet at its base. Construction was begun by tying together, about a foot from their tops, three slender poles about sixteen feet long. The butts of these were spread out to the diameter desired and firmly imbedded in the ground. Then a dozen or more poles were laid against the three basic poles to form a circle to support the slabs of bark. Where the entrance was to be, a short length of pole was fastened horizontally across two of the poles about three feet from the ground.

From this point on, the slabs of bark were put in place against the supporting poles, beginning at the bottom. When these had been placed around the entire base of the house, the second layer of bark was placed so as to overlap the tops of the first, and, when necessary, a third layer was placed to overlap the tops of the second. Holes that remained were covered with short pieces of bark and care was taken to leave a small smoke-hole at the very top of the structure. Stones were placed against the outside circle of bark. As the work progressed Calistro would often in a very gentle voice sing snatches of Cahuilla song, but in his diffident way he would refuse to tell us about them. Also, under his direction, a tiny fire pit, lined with stones, was dug in the center of the floor. Beds, filled with pine needles and oak leaves, were made of small diameter logs and each of the three sides of the "kish," and a low table, made from a slab of wood, was placed close by the fire pit.

In the old days, Calistro told us, a piece of hide served to close the doorway against storms, and strips of rawhide were used to tie the tops of the first three poles together as well as to fasten the short piece of pole of the doorway in proper position. We had to use stout wire and a piece of canvas.

When the structure was finished, Calistro expressed satisfaction with everything except the doorway, which he felt was far too large.

During the rain and snow storms that have beat upon the bark house in the years since it was erected, we have been astonished how weatherproof it has proved to be. We were particularly surprised to find that the bark remained in place even during heavy windstorms. This may have been due to the fact that Calistro insisted that it be erected in a sheltered spot facing either east or southeast.

As we were working, Calistro supervised construction from a comfortable chair in the shade. Midway he called us over to assure us: "This is how it should be . . . the white man does all the real work and the old Indian sits in comfort in the shade."

Lotus scoparius (Nutt.) Ottley (Leguminosae)

Common Name: *Deerweed*   Cahuilla Name: *kiwat*

This bushy annual is found on dry slopes and fans, especially after burns below 5,000 feet. Mrs. Alice Lopez recalled that the plant was used as a material in house construction.

Lupinus L. (Leguminosae)

Common Name: *Lupine*   Cahuilla Name: *See Below*

No Cahuilla of today recall any practical uses for the various species of lupine found in Cahuilla territory. Mrs. Gertrude Chorre, a Luiseño from La Jolla Reservation, has recalled that lupine leaves were sometimes eaten by her people. One species, *L. hirsuitissimus* Benth., was identified by Mrs. Alice Lopez as having the Cahuilla name of *samat iwyak* or "thorny bush." Another lupine species, unidentified, was referred to as *tamit meh'a* or "rays of the sun" by Cinciona Lubo.

Lycium L. (Solanaceae)

Common Name: *Box-Thorn*   Cahuilla Name: N.D.

The Cahuilla are known to have eaten the berries of at least two species of *Lycium* found in their territory, *L. Fremontii* Gray and *L. andersonii* Gray. The first species is common to alkaline places on the Colorado Desert below 1,500 feet. The latter species may be found on dry stony hills and mesas below 6,000 feet, particularly in the pinyon-juniper areas.

Both shrubs produce berries in considerable abundance between May and August. The berries were eaten fresh or preserved by drying. The dried berry was boiled into a mush or ground into flour and mixed with water. According to Cahuilla interviewed, all shrubs of these species do not produce edible fruit. The shrubs do produce with considerable regularity, however, and were considered a significant item in the Cahuilla diet.

Malva L. (Malvaceae)

Common Name: *Mallow, Cheeses,*   Cahuilla Name: *N.D.*
Cheeseweed

Several species of this common introduced herb probably can be found in Cahuilla territory. The seeds of the mallow, available from February to late summer in some places, were eaten fresh by the Cahuilla. The seeds were said to be a pleasant condiment-like food.

Marrubium vulgare L. (Labiatae)

Common Name: *Horehound*   Cahuilla Name: *N.D.*

This introduced plant was boiled whole to make an infusion for flushing the kidneys. Mrs. Alice Lopez referred to the plant as *chungal*, which is a descriptive term meaning "sticky to the touch." She said there was a Cahuilla name for the plant, but she could not recall it.

Matricaria matricarioides (Less.) Porter (Compositae)

Common Name: *Pineapple Weed*   Cahuilla Name: *N.D.*

Mrs. Alice Lopez and Cinciona Lubo remember this common weed as one from which an infusion was made to settle upset stomach or cure diarrhea and colic.

Medicago hispida Gaertn. (Leguminosae)

Common Name: *Bur-Clover*   Cahuilla Name: *N.D.*

The bur-clover, a European native, is found frequently in grassy places in Cahuilla territory, particularly in the Upper Sonoran zone. The seeds of the plant were harvested in the spring and early summer. They were parched, ground, and made into a mush.

Mentzelia L. (Lorasaceae)

Common Name: *Blazing-Star*   Cahuilla Name: *N.D.*

A dozen or more *Mentzelia* species may be found growing in the Cahuilla environment. These include *M. involucrata* Wats., found in the Coachella Valley; *M. Veatchiana* Kell., *M. Puberula* J. Darl., and *M. albicaulis* Dougl. ex Hook. (see Munz, 1965, for other species). Various species may be found from both the lower to the upper Sonoran life zones. These plants provided edible seeds from February until October, depending upon species and location. The seeds were parched and ground into flour for use as a mush.

Monardella villosa Benth. (Labiatae)

Common Name: *Coyote-Mint, Pennyroyal*   Cahuilla Name. N.D.

Romero (1954:8) reported that the Cahuilla used this perennial herb for the relief of stomachache, a usage confirmed by Cinciona Lubo. Mrs. Lubo said the leaves were made into a tea for this purpose. Romero recorded the Cahuilla name of the plant as *tah-lis-wet*, which would appear to be a rendering of *taxliswet* or "human being." None of the Cahuilla interviewed by the authors recall this word as a plant name. The neighboring Luiseño Indians use the name *havawut* for coyote-mint (Sparkman 1908:211).

Montia perfoliata (Donn.) Howell (Portulacaceae)

Common Name: *Miner's Lettuce*   Cahuilla Name: *palsingat*

Miner's lettuce is commonly found in moist places, especially along streams and in wet valleys, as high as 6,000 feet. The plant was eaten fresh or boiled as a green. The Cahuilla name given above came from Mrs. Katherine Saubel. (Romero 1954:62) recorded the plant's name as *lahchumeek*, but no Cahuilla interviewed recognized the term. Another species of the plant, M. *spathulata* (Dougl.) Howell is also found in Cahuilla territory, and may have been used as a green also. Both species would have been available for gathering in the early spring.

Muhlenbergia rigens (Benth.) Hitchc. (Gramineae)

Common Name: *Deer-Grass*   Cahuilla Name: *suul*

The stalk of this plant (formerly *Epicampes rigens californica*) was widely used in basketmaking by California Indians and it was one of the primary plants employed by the Cahuilla in basketmaking. The plant grows in mountainous areas and was often picked in the summer by Cahuilla men who encountered it while hunting. The stalk of the plant was used as the horizontal or foundation elements around which the coils were wrapped.

Locations where the plants grew were often claimed as private property by individual Cahuilla. Often locations were kept secret to prevent random collection by others. The plants are still sought after today by older Cahuilla, but are increasingly difficult to find.

Nasturtium officinale R. Br. (Cruciferae)

Common Name: *Water-Cress*   Cahuilla Name: *pangasamat* [?]

The water-cress is common to wet areas up to 8,000 feet throughout much of Cahuilla territory. This aquatic perennial was eaten fresh in the spring or cooked like spinach and flavored prior to eating. According to Mrs. Alice Lopez, the tops had a peppery taste. Mrs. Jane Penn stated that water-cress often was mixed with less flavorful greens into a salad.

The Cahuilla name given above was recorded by Romero (1954:65), but we have been unable to confirm it. The name is said to mean "grass in the water." Romero reported that water-cress eaten each morning with salt would cure liver ailments within two months. The patient was expected to eat no other food before noon. The plant also is said to have been effective in treating low blood pressure.

Nicotiana L. (Solanaceae)

Common Name: *Tobacco*   Cahuilla Name: *pivat*

Tobacco is the sacred plant of the Cahuilla, and its usage probably is more ancient than that of its hallucinogenic relative of the Solanaceae family, datura. Tobacco was one of the first plants created by the god Mukat, who drew both the plant and a pipe from out of his heart. Seeking a means to light his pipe, Mukat then created the Sun, but it escaped from his grasp, and he was forced instead to light his pipe with the Western Light (Strong 1929:132). Tobacco figures prominently throughout Cahuilla oral literature, where it is often associated with power, curing, gaming, the human soul, and many other activities and concepts.

At least four species of *Nicotiana* are indigenous to Cahuilla territory: *N. trigonphylla* Dunal, *N. attenuata* Torr., *N. Bigelovii* (Torr.) Wats., and *N. Clevelandii* Gray. The introduced South American species *N. glauca* Grah. is called tree tobacco. Barrows (1900:74-75) was familiar only with the smoking of *N. attenuata*, although he did note that Palmer (1878:650) had spoken also of the use of *N. trigonophylla* and *N. Bigelovii* among most of the Indian groups of southern California. Modern Cahuilla have confirmed the use of *N. trigonophylla* and *N. glauca* as well. *N. Clevelandii* is an occasional plant, often found in dry barrancas after a single rain. It seems probable that this species too was used by the Cahuilla.

Barrows (1900:75) reported that the white man's tobacco was called *pivat*, and that the Cahuilla tobacco *N. attenuata* was referred to as *pivat-isil* or "coyote's tobacco." Several Cahuilla interviewed by the authors disagreed, however, insisting that *pivat* referred to a specimen of *N. trigonophylla* which the authors showed them and that *isil pivat* (note the reversal of the two words) was the correct designation for the non-native species of *N. glauca*. Possibly all indigenous species of tobacco were known as *pivat* to distinguish the traditional tobacco from the plant introduction. Nevertheless, the authors have not been able to satisfactorily resolve this point with any degree of certainty. None of these plants, of course, is the cultivated, commercially-produced tobacco, *N. tobacum.*

Several and possibly all of the tobacco species used by the Cahuilla contain alkaloids which are highly toxic, affecting both the central and peripheral nervous system and increasing the activity of the secreting glands. *N. glauca* has been the cause of illness in children sucking the flowers, and death when the leaves were used in a green salad (Hardin and Arena 1969:116). *N. trigonophylla* caused poisoning and one death in a California family that ate its leaves as boiled greens (Hardin and Arena 1969:116).

Among the Cahuilla, tobacco was chewed, smoked, or used in a drinkable decoction depending upon the purpose for which it was intended. For smoking, several varieties of pipes were used. Stone and clay pipes, occasionally with reed stems, were most common. These pipes, owned by the men who used them, were for both day-to-day and special uses. Certain pipes sacred to ritual were kept in the Cahuilla ceremonial bundle (*maiswat*) and brought forth only for community rituals. Other sacred pipes were owned and used by shamans (*puvulam*) for curing and other shamanistic activities. Pipes were often decorated by incising the clay or stone. Two common shapes of pipes were the tubular one without a handle, usually smoked in a prone position or with the head tilted back, and a smaller pipe, usually made of clay, which had a handle on the bottom. Sometimes pipes were made of cane with a reed stem. Drucker (1937:25) even recorded the use of a double bowl pottery pipe.

Ritual Uses of Tobacco.—Tobacco was an integral part of every ritual, just as it was a significant event at the time of creation when the god Mukat created tobacco and smoked it ceremonially with his brother god Temaiyawit (Strong 1929:132; Hooper 1920:318). The account of this first use of tobacco is an important part of the song texts sung at the most important Cahuilla rituals. In fact, an abundance of tobacco was and still is as important to ritual etiquette as having an abundant amount of food for guests, virtues that would later contribute to the successful flight of the soul to the land of the dead.

Before a ritual was conducted, tobacco was smoked by the ritual leaders and shamans and the smoke was blown in the sacred directions: north, east, west, south, and up or center. This helped to clear the area of any malevolent force which might interfere with the ritual. Throughout ceremonies—especially those honoring the recent dead (*nukil*) everyone was obliged to smoke tobacco—as they are even today. At funerals, smoking serves to concentrate power that will aid the dead in their arduous journey to the other world.

The ceremonial bundle—that most sacred object of the Cahuilla which contained ritual objects and an emanation of the creative spirit *'amna'*—was regularly "fed" sprinklings of tobacco.

Tobacco Use by Shamans.—Since tobacco was intimately associated with power and to some degree with ecstatic experience, it was a basic part of the shaman's equipment. Each shaman had his special pipes which he utilized in curing, dreaming, and his various other activities. Shamans usually owned their own plots of tobacco, which they gardened and processed. Cultivation of tobacco was a form of incipient agriculture among the Cahuilla.

Shamans were expected to use their special powers, made possible through their intimate connections with supernatural beings, in aiding the community. Among their concerns were the control of rain, crop production, divining, and the general health of the community. Tobacco was utilized in several of these activities. In order to influence crop productivity, for example, shamans would place a small portion of tobacco in the palm of their hands during public rituals and cause miniature food-producing plants to grow from the tobacco. Miss Victoria Wiereck recalled witnessing a miniature oak tree grow out of the tobacco in the palm of a *puul* on one occasion. Such a demonstration of great power guaranteed supernatural aid in protecting a particular food supply so that it would be available for gathering.

Tobacco was also employed in shamanistic ceremonies at community gatherings to drive away malevolent powers. Several nights of singing and dancing often accompanied power demonstrations in which tobacco was smoked and blown in the proper directions to make it possible to discover anyone in attendance who had ill intentions toward anyone else and to disperse all evil presences.

The shaman frequently used tobacco in curing, since its power helped to alleviate the cause of a sickness, whether it was caused by a lost soul or an introduced disease object placed in the patient's body by a malevolent person. Even when sickness stemmed from supernatural forces, blowing smoke over the patient helped concentrate the power of the curer, purified the patient, and assisted in eliminating the sickness.

Tobacco smoking helped the shaman in establishing rapport with supernatural helpers and aided him in "dreaming" as he sought to divine the origin of illnesses or future events. Frequently, smoke was used to purify or "fix" things that had gone wrong—for example, a piece of equipment that had been touched with menstrual blood or food which had been "poisoned" by malevolent forces.

Medical Uses of Tobacco.—Tobacco was used in a number of medicinal remedies. A water solution of tobacco served as an emetic to induce vomiting. Leaves of tobacco were employed as poultices to heal cuts, bruises, swellings, and other wounds. To alleviate earaches, tobacco smoke was blown into the ear, which was then covered with a warm pad. The smoke was said to "go in all the way and relieve the pain."

Romero (1954:61) also recorded the use of tobacco in curing scrofula and rheumatism. Leaves were steamed and applied externally to the lymph gland areas of the neck as a cure for scrofula or other forms of throat inflammation. In the case of rheumatism, tobacco leaves were placed on hot rocks in the sweathouse and the patient inhaled the steam. Patients also sometimes inhaled tobacco steam as a cure for nasal congestion.

Other Uses of Tobacco.—There were no special restrictions on who could smoke tobacco, and older men and women in particular smoked it almost daily when the plant was available. Tobacco was considered a relaxing euphoriant as well as a medicinal and ritualistic plant. Men often smoked tobacco as part of a hunting ritual. In particular, it was frequently used by hunters who had gone through a fasting ritual prior to the hunt for the purpose of allevating hunger. According to Lucille Hooper (1920:348), women drank a weak tea made from tobacco at the time of menstruation to keep the body free from unpleasant odors.

Tobacco was ritually smoked by individuals embarking on any sort of important activity, particularly a journey. The traveler would blow smoke in the direction he intended to go in order to clear away all danger and ensure that he would have the blessing of his spiritual guides.

Nolina Bigelovii (Torr.) Wats. (Agavaceae)

Common Name: *Nolina*   Cahuilla Name: *kuku'ul*

This yucca-like perennial grows on dry slopes below 3,000 feet. The plant usually blossoms from May to June. The stalk was baked in a rock-lined roasting pit, where it was covered with sand and left overnight. The taste was said to be somewhat bitter. Barrows (1900:59) was probably referring to this plant when he spoke of a "different variety of yucca" of an unknown species, which he said the Cahuilla called "*ku-ku-ul.*"

Oenothera clavaeformis Torr. & Frem. (Onagraceae)

Common Name: *Evening Primrose,*   Cahuilla Name: *tesavel*
               Desert Primrose

Mrs. Alice Lopez recalled that this plant grew in abundance near the Narbonne ranch, where she went with her mother and a "lot of other people" as a small child to gather the plant, which was used as a green. An added delicacy was the frequent presence on the plant of the caterpillar of the white line sphinx moth, *Celerio lineata* (Fabr.), which the Cahuilla called *piyakhtem.* These caterpillars were a favorite food that Cahuilla have compared in taste to pork rinds. The heads of the caterpillars were chopped off, the insides were cleaned out, and they were boiled, parboiled, or dried in the sun.

Olneya Tesota Gray (Leguminosae)

Common Name: *Desert-Ironwood,*   Cahuilla Name: N.D.
               Ironwood

This grayish tree with scaly bark may be found in desert washes below the 2,000 foot level, often in association with mesquite and palo verde. The pods and seeds of the tree were gathered from May to June and roasted and ground into flour. The desert-ironwood was also used to fashion implements requiring extreme hardness, such as throwing sticks and clubs. It was also considered an excellent firewood.

Opuntia acanthocarpa Engelm. & Bigel. (Cactaceae)

Common Name: *Buckhorn Cholla*   Cahuilla Name: *mutal*

Although Barrows (1900:68) gave the name *mutal* to what appears to have been another species of *Opuntia*, which he was unable to identify, Cahuilla interviewed by the authors agreed that *mutal* was the buckhorn cholla. This cactus is found on dry mesas and slopes below 4,500 feet. The fruit was gathered in the spring and eaten fresh or dried for storage. Ashes of the stems of this cactus were applied to cuts and burns to facilitate healing.

Opuntia basilaris Engelm. & Bigel. (Cactaceae)

Common Name: *Beavertail*   Cahuilla Name: *manal*

Often found on fans and dry benches below 6,000 feet, beaver-tail grows in abundance in many areas of the desert, sometimes in stands covering several acres. During its season between March and June, beavertail was plentiful and considered one of the most desir-able of Cahuilla foods.

Barrows (1900:67) wrote as follows about this cactus: "It is one of the small varieties and has a tender slate-covered stem in flat joints. The young fruit in early summer is full of sweetness. These buds are collected in baskets, being easily broken off with a stick. The short, spare spines are wholly brushed off with a bunch of grass or a handful of brush twigs. The buds are cooked or steamed with hot stones in a pit for twelve hours or more."

Up to three of four buds usually sprout from the upper edge of each joint, making several dozen buds available on each plant. After the buds were cooked, they could be eaten immediately or dried for indefinite storage. When young and tender, the joints also served as food. They were cut into small pieces, boiled in water, and mixed with other foods or eaten as greens. The rather large seeds of the beavertail were ground into an edible mush.

Opuntia Bigelovii Engelm. (Cactaceae)

Common Name: *Jumping Cholla,*   Cahuilla Name: *chukal*
 Ball Cholla

The notorious jumping cholla is common to desert fans, benches, and lower slopes below 3,000 feet. Barrows (1900:68) described it as ". . . a furry cactus with round jointed stems two or three feet high. It is light brown in color and grows in communities, sometimes covering a rocky canyon site for a half mile to the exclusion of almost everything else. It throws off extremely disagreeable balls of spines which fasten in a horse's fetlocks and give instant trouble." Despite the inconvenience of its spines, the buds of the jumping cholla were gathered regularly by the Cahuilla and prepared and preserved in a fashion similar to the beavertail. The buds were available from late April to June.

Opuntia megacantha Salm-Dyck. (Cactaceae)

Common Name: *Tuna*   Cahuilla Name: *navet*

The tree-like tuna cactus was a significant food source of the Cahuilla. The plant was usually found on dry slopes or walls between the 3,000 and 5,000 foot level. Barrows (1900:67) reported that the tuna cactus did not grow on the Cahuilla Reservation of the Santa Rosa Mountains, and that these Indians sometimes obtained tuna from groups in the Coachella Valley. He noted that the bud-like fruit was called "*na-vit-yu-lu-ku* or 'the little heads of the cactus.'" An introduced plant from tropical America, the tuna probably was quickly adopted into the Cahuilla economic system.

The pads of the tuna were gathered when young, anywhere from May until August, and diced. The pieces were then boiled and eaten or dried and stored for later use. Large quantities of tuna could be gathered and stored for lean periods. The plant also produces a profusion of edible buds from two to four inches in length, and a small stand of these cacti can produce hundreds of edible buds in a season. The buds were picked and eaten fresh or dried for future use. Gathering was a difficult process, since the buds are covered with innumerable tiny spines, which easily stick to the fingers or may even blow into a picker's eyes if there is a wind. The buds were therefore usually picked in early morning when they were damp with dew. This caused the spines to lie flat and made them less troublesome. The fruit was also cool in the morning and refreshing to eat, making an excellent early morning meal. During preparation, the fruit was held in such a manner that the ends could be severed. An incision was made lengthwise in the fruit and the outer skin peeled off. If this operation is performed carelessly, the spines fall into the flesh of the fruit, later sticking in the mouth of the eater. The Cahuilla were able to peel the fruit with astonishing speed and skill.

When boiled, the fruit was said to make an excellent purgative. Some Cahuilla still use it as a cure for constipation. According to Romero (1954:16) and several Cahuilla interviewed by the authors, plugs were made from the plant for inserting into wounds as healing agents.

Opuntia occidentalis Engelm. & Bigel. (Cactaceae)

Common Name: *Prickly-Pear*   Cahuilla Name: *qexe'yily*

At least three varieties of the prickly-pear may be found in Cahuilla territory. Probably the fruit of all these varieties were eaten by the Cahuilla, but the variety identified for the authors was *megacarpa* (Griffiths) Munz. This variety is found on dry slopes at the edge of the desert between the 3,000 and 7,000 foot levels. The edible fruit was ready for harvesting in May and June. The joints also were diced and eaten.

Opuntia ramosissima Engelm. (Cactaceae)

Common Name: *Pencil Cactus*   Cahuilla Name: *wival*

The pencil cactus is common to dry washes, slopes, and mesas of the desert below 4,000 feet. The edible fruits were gathered between April and May and eaten fresh or dried for later use. The stalk, with thorns removed, was boiled into a soup or preserved for future use by drying.

Orobanche ludoviciana Nutt. var. Cooperi (Gray) G. Beck.
   (Orobanchaceae)

Common Name: *Broom Rape*   Cahuilla Name: *meslam*

Broom rape, a parasitic plant with thick stem, a root clump, and no green parts, is found in desert areas up to 4,000 feet. The plant lies mostly below ground, thrusting through the soil when about to bloom.

Barrows (1900:66) described the plant and its usage as follows: "It has large succulent roots, yellow or white, and in springtime, before the plant blossoms, and while the roots are young and tender, they are dug up and roasted in the coals for food." Patencio (1943:99) called the plant *mesalam* and wrote: "Down the Chino wash, across the boulevard, a plant grew stalks like the asparagus, but had a root like sweet potatoes. This plant was gathered in the spring by the Indians."

Use of the plant is only scarcely remembered today, and we were unable to ascertain the extent to which the roots were used as a food. Older Cahuilla did recall that it was gathered from about April to June and peeled prior to eating.

Paeonia Brownii Dougl. ex Hook. (Paeoniaceae)

Common Name: *Peony, Western Peony*   Cahuilla Name: *N.D.*

Palmer (1878:65) reported that the roots of the peony, a perennial herb, were used by the Indian groups of southern California in curing colds, sore throats, and chest pains. In this regard, he wrote: "It is mealy and tastes somewhat like licorice. After being reduced to powder, it is either taken in that form internally or made into a decoction." The Surprise Valley Paiute also used the plant medicinally, making a cough medicine called *butipi* from the seeds.

Palafoxia linearis (Cav.) Lag. (Compositae)

Common Name: *Spanish Needles*   Cahuilla Name: *tesqal*

This annual herb, found on sandy flats and in washes of the Mohave and Colorado deserts up to 2,500 feet, was used by the Cahuilla as a source of yellow dye.

Panicum Urvilleanum Kunth. (Gramineae)

Common Name: *Panic Grass*   Cahuilla Name: *sangval*

Several species of panic grass grow in Cahuilla territory, and probably the seeds of all of them were used for food. This particular species, however, was pointed out to the authors as having been used to make gruel. The seeds of the plant were singed to remove hair and then boiled for several hours.

Saunders (1914:136) reported the use of panic grass among the Cahuilla and recorded the name of *songwall*. According to Paul Chase, former curator of Bowers Museum, Santa Ana, an olla found by hikers near Palm Springs in 1969 was filled with the seeds of *Panicum Urvilleanum*.

Earle and Jones (1962:224) analyzed one panic grass species and found that it contained 4.0% ash, 15.0% protein, 5.8% oil, some starch, and no tannin or alkaloids. Saunders (1919:136) suggested that its nutritional value was similar to that of millet.

Penstemon centranthifolius Benth. (Scrophulariaceae)

Common Name: *Scarlet Bugler*   Cahuilla Name: *tuchilychungva*

Modern Cahuilla often gather the scarlet bugler for use as decoration at funerals or church affairs, but whether it was similarly used for ornamental purposes in earlier times is not known. The name *tuchilychungva* means "hummingbird's kiss," a reference probably to the attractiveness of the flowers to this bird. Sometimes the plant is called *pisily* or "sweet."

Perezia microcephala (DC.) Gray (Compositae)

Common Name: *Scapollote*   Cahuilla Name: *"ha-bak-a-ba"*

Barrows (1900:78) noted that this perennial herb, two or three feet high, with thin oblong leaves and purple flowers, was called *"ha-bak-a-ba"* by the Cahuilla, but present-day Cahuilla are unfamiliar with the word. According to Barrows, the plant was used as a decoction that provided a swift remedy for constipation.

Perideridia Gairdneri (H. & A.) Math. (Umbelliferae)

Common Name: *None in Munz*   Cahuilla Name: *N.D.*

According to Saunders (1919:13), the tuberous roots of this herb were eaten raw or cooked. Munz (1965:1012) also reported that roots of the herb were an important food source for Indians in the Southwest.

Phaseolus L. (Leguminosae)

Common Name: *Beans*   Cahuilla Name: *tevinymalem*

At least two species of beans were cultivated by the Cahuilla early in the historic period: tepary beans (*Phaseolus acutifolius* L.) and the kidney bean (*Phaseolus vulgaris* L.). The tepary bean was grown aboriginally by the Colorado River tribes to the east of the Cahuilla (Castetter and Bell 1951:107), and may have reached the Cahuilla prior to Spanish contact. Although the kidney bean was aboriginal in the Southwest, having been grown by the Hohokam, no evidence exists that it had reached the Yuman tribes of the Colorado River prior to the arrival of Europeans (Castetter and Bell 1951: 107-108).

Strong (1928:38) interviewed elderly Cahuilla born in desert villages of the Coachella Valley as far back as the 1850's, and within their memory ". . . both corn and wheat were raised . . . and doubtless other vegetables such as melons, beans, and squash." Barrows (1900:72-73) reported that beans had been cultivated among the Cahuilla for a "great many years" and that it was "easier to imagine that the knowledge of agriculture with the seed of corn, squash, and bean came to them long ago across the desert, than that they learned of these things only in this century from the Spaniards." (See Appendix.)

According to Mrs. Katherine Saubel, the word *tevinymalem* for beans was applied to "the very old beans" among her people, which she described as small, dark brown beans, oblong and slightly round. This description would best apply to a variety of the tepary beans grown aboriginally along the Colorado River. The larger frijole or kidney bean, popular among the Mexicans, was known as *huul* (singular) or *huuluam* (plural) and is a Spanish loan word derived from *frijole*. Francisco Patencio (1943:25) reported that *Ta va my lum* was the name of the beans that grew out of the fingers of the dead god Mukat in the Cahuilla creation myth. J. P. Harrington (unpublished) in his field notes recorded *tevasmal* (singular) and *tevasmalem* (plural) as the words that his informant, Adan Castillo, succeeded in recalling for beans. Castillo believed that *tevas* meant "little round things."

Phoradendron Nutt. (Loranthaceae)

Common Name: *Mistletoe*   Cahuilla Name: *chayal*

Two species of mistletoe were used by the Cahuilla: juniper mistletoe (*P. ligatum* [Trel.] Fosb.) and desert mistletoe (*P. californicum* Nutt.) The former is found upon the juniper tree, and the latter parasitic plant is found mostly on the catclaw and mesquite.

Barrows (1900:80) reported that the berries of the juniper mistletoe were sometimes eaten fresh, but more frequently pounded into a flour which was sprinkled into wounds to aid healing. Cinciona Lubo also recalled that the powder, mixed with water, was sometimes used to bathe sore or infected eyes.

The berries of the desert mistletoe were available from November to April. They were ground and mixed with a small amount of ashes (to counteract inherent viscidity) and boiled in a pot. The taste was said to be somewhat sweet. Curtis (1926:24) noted that the berries were used like a dessert as an occasional treat. Hardin and Arena (1969:76) warned that the berries are toxic and may cause acute stomach and intestinal irritation, accompanied by diarrhea and a slow pulse.

Leaves of the desert mistletoe were used as a source of dye for basket weeds. Basket weeds were boiled in a mixture of the leaves, which permanently dyed them black. The name *chayal* refers to the berries and not the plant itself. Mrs. Alice Lopez recalled that the leaves were sometimes boiled into a tea, which may have had a medicinal use.

Phragmites communis Trin. var. **Berlandieri** (Fourn.) Fern. (Gramineae)

Common Name: *Common Reed*   Cahuilla Name: *pakhal*

Common reed grows in abundance in canelike thickets in many wet areas below 5,000 feet both on the desert and in other scattered localities throughout Cahuilla territory.

Barrows (1900: 37, 47, 49, 50) reported a variety of uses for common reed. In the mountains, it was often used as a thatching material for house construction. For making cordage, Barrows (1900:47) said: "These stems are soaked in water and then the bark is easily removed, a layer of soft, silky, yellowish brown fibers. It is twisted into a beautiful and very strong cordage." The cordage was used in weaving carrying nets and in making hammocks for babies.

The culm of the plant, sometimes five to ten feet high, provided a shaft for arrows (Barrows 1900:50) of the two-piece variety. The head was usually made of mesquite or greasewood inserted into the hollow reed. The common reed also provided material for a flute usually played by men. A splint was also made from the common reed in treating broken limbs. A specimen collected by A. L. Kroeber is at the Lowie Museum.

The giant reed (*Arundo Donax* L.), an introduced plant from Europe, was used for many of the same purposes as the native common reed. Both plants are called *pakhal* by Cahuilla today. The giant reed spread rapidly throughout southern California, particularly along the Los Angeles River, where it was so abundant in the 1820's that Mexicans used it in roofing their houses.

Pinus L. (Pinaceae)

Common Name: *Pine*   Cahuilla Name: *wexet*

Several species of these evergreen trees may be found in the various mountainous areas of Cahuilla territory, including the limber pine (*P. flexilis* James), the sugar pine (*P. Lambertiana* Dougl.), the lodgepole pine (*P. Murrayana* Grev. & Balf.), the ponderosa pine (*P. ponderosa* Dougl. ex. P. & C. Lawson), and the knobcone pine (*P. attenuata* Lemmon.). All of these may have had their occasional uses, but the two most important species to the Cahuilla were the low-growing nut or pinyon pines, *P. monophylla* Torr. & Frem., and *P. quadrifolia* Parl. ex Sudw.

The name *wexet* encompassed all pines, especially the larger pines, but the name *tevat* was applied to the pinyons. The affix *wik* was often added for *P. quadrifolia*, making the name *tevatwik* or literally "fat pinyon." *P. monophylla*, commonly called the one-leaved pinyon, may be found on dry rocky slopes and ridges from 3,500 to 9,000 feet. The four-leaved pinyon, *P. quadrifolia*, was common on dry slopes from 2,500 to 8,000 feet. The area around Santa Rosa Reservation was a typical pinyon gathering area. Abundant stands of pinyon may be found on Toro Mountain near Coyote Canyon, on the peaks of Black Hill and Sheep Mountain, within the upper reaches of Deep Cayons, and across the plains of Pinyon Flat and Little Pinyon Flat. Pinyon trees are usually found mixed with stands of juniper, scrub oak, and other shrubs. A third pinyon species, *P. edulus* Engelm., is common to the Little San Bernardino Mountains, where it was undoubtedly utilized by some Cahuilla groups.

Areas where pinyon nuts were found in abundance during good years were close to most Cahuilla villages. Pinyon was an important food source, but the trees were unreliable, sometimes producing a rich harvest and at other times little or no crop. Pinyon nuts were harvested in early summer or late August. According to Calistro Tortes, a ceremonial eating of a few pinyons was mandatory prior to harvesting. No settlement sites have been located in the pinyon groves. The Cahuilla journeying to them never erected permanent houses, since harvesting was carried out in a two- to six-day period. Simple shelters were used or the gatherers simply camped in the open.

Generally, except for the very young or the old and infirm, the entire family participated in the pinyon harvest. Trail complexes into the pinyon-gathering areas are even today well distinguished. There was a significant lack of ownership specificity for pinyon because of the undependability of the harvest. Political boundaries were apparently adjusted to give as many groups as possible access to the nuts, and Cahuilla came from several political groups to such areas as Pinyon Flats, which appeared to be an area on which many groups bordered. In this area, people from the Colorado Desert, Palm Springs, and the Santa Rosa Mountains all met during harvest.

During the early summer harvest, the cones—not yet fully ripened—were knocked from the trees with a hooked pole. These green cones were then placed in a pile and a fire was set over them or they were baked in a pit. The heat prematurely opened the cones, so that the pinyon nuts could be collected before their natural opening when competition with animals and birds was acute.

The month of August was the usual time for harvest. The cones were usually ready to dislodge from the trees. Since the nuts were rapidly eaten by rodents and birds, timing was important in acquiring the nuts. Cahuilla usually tried to arrive at the pinyon stands just before the heaviest drop occurred. Cones were easily knocked from the trees by shaking or by poles. Cones were collected in baskets by women and placed in a bed or pit of hot coals. Heating melted the resin which caused the nuts to cling to the cones and the nuts were prematurely released. Archaelogical evidence of the roasting process may frequently be found in pinyon areas in the form of a small ring of rocks. In regard to the roasting of the pinyons, Palmer (1878:595) wrote: "After a few stirrings, they are sufficiently parched to render the hull brittle, so as to be easily removed, while the oil in the kernal is set free. By this process the kernel is rendered more digestible and will keep for a longer time."

After roasting, shells of the nuts were removed by placing them in a metate and rolling a mano over them. The nut was then winnowed in a basket prior to eating and cooking. Pinyon nuts were eaten whole or ground and made into mush. A palatable drink was made by mixing a small amount of ground nuts with water. The presence of metates at pinyon gathering sites suggests that mush was sometimes made while the group was camped in a pinyon grove.

The cooked unshelled nuts were brought back to villages and stored. One Cahuilla interviewed by the authors recalled gathering two one-hundred pound sacks of pinyon nuts for a single family. Nuts were stored in ollas or underground caches. Calistro Tortes of the Santa Rosa Reservation recalled storing roasted nuts in sacks in holes dug in the ground. Pinyon nuts were one of the few foods fed to babies in lieu of a natural milk diet.

Zigmond (1941:32) reported pinyon nuts are high in fat content (35.4% whole and 60.7% hulled), highly energizing, and very close in caloric value per pound to almonds, brazil nuts, peanut butter, coconuts, and bacon. He also suggested that they were probably deficient in certain important nutritive constituents. He concluded: "If eaten in considerable amount, part of the fat content would be converted into body fat which could be utilized as energy in times of food scarcity. With compensating additional food sources and sunshine, one might well find pine nuts a satisfactory food staple. Indeed, to the uncertain nutritional conditions typical of the aboriginal Great Basin, a food supplying an excessive energy potential which might be consumed in lean times would seem an altogether sound adjustment.

Because of its high combustibility, pinyon wood was sought as firewood, especially for kindling. Small slivers of the pine trees were used for torches and so flame could be carried from one site to another. The hot flame that gives off a pleasant odor increases its attraction as a firewood.

Pinyon needles and those of the larger pines were used as basketry material. In pine needle baskets, the smaller roots of pine were used to bind the needles together. Pine pitch provided an adhesive for many objects, including mending of pottery and baskets and attaching arrowpoints to shafts. According to Cinciona Lubo, pine pitch was used as a face cream by girls to prevent sunburn. The bark of pine trees was a desired roofing material in house construction.

Pinyon nuts were an important trade item of the Cahuilla. They were traded among themselves and with the neighboring Luiseño and Diegueño. Cahuilla groups that lacked pinyon stands in their area often made arrangements with outside Indian groups to gather in their territory. The Wanikik Cahuilla of San Gorgonio Pass, for example, regularly paid a fee to the Serrano Indians to gather nuts in their groves in the San Bernardino Mountains.

Platanus racemosa Nutt.

Common Name: *Sycamore, Plane-Tree*   Cahuilla Name: *sivily*

The sycamore is common to streambeds and watercourses below 4,000 feet in many areas of Cahuilla territory. Limbs and branches of the sycamore were employed in house construction. J. P. Harrington (unpublished) was told by Adan Castillo that wooden bowls were sometimes made from sycamore wood. They were shaped with a broken rock, seasoned in water, and greased with meat or oil to prevent splitting.

Pluchea sericea (Nutt.) Cov. (Compositae)

Common Name: *Arrowweed,*   Cahuilla Name: *hangal*
             Marsh Fleabane

This slender willowlike shrub grows abundantly in wet places such as along streambeds in both the lower and upper Sonoran life zones. Found frequently in alkaline soils, arrowweed may cluster in dense thickets as high as ten feet. The plant flowers in early spring from February until May, depending on locality.

The roots of the young plant were gathered for roasting and eaten. Since some young plants were always available, the roots remained a constant source of food.

The long, slender, pliable stems with their leaves were gathered in quantity for house construction. As a roofing material, the plant provided an excellent protective roofing material. The flexibility of the stems also made it possible to interweave them with stronger materials in the walls of houses as part of a wattle and daub construction. The result was an almost solid wall mass that was impervious to rains and wind. Arrowhead was also used in construction of ramadas, windbreaks, fences, and granaries.

The plant was used also by the Cahuilla and many other Indian groups of the Southwest in making arrows. A shaft of arrowhead was cut to the desired length and moistened. It was then placed in the groove of a heatened stone arrow straightener. The shaft was then worked back and forth until natural curvatures were straightened out.

Populus Fremontii Wats. (Salicaceae)

Common Name: *Cottonwood,*   Cahuilla Name: *lavalvanat*
Aspen, Popular

The cottonwood tree is found in moist places below 6,500 feet. The wood provided material for manufactured tools of various kinds and excellent firewood. The bark provided a fibrous material that could be used for a variety of purposes.

One of the unusual uses of the cottonwood was in making wooden mortars. The trunk of a cottonwood tree was cut into two and one-half foot lengths. Part of the center of the trunk was then removed to create a mortar in which soft foods such as mesquite could be ground.

Cottonwood also had a number of medicinal uses. The leaves and bark were boiled into a poultice to relieve swelling caused by muscle strain. The injured limb was wrapped with the poultice or soaked in a solution of hot water, bark, and leaves. Cinciona Lubo recalled that her family used cottonwood solutions in treating cuts. The cut area was washed in warm water and covered with a solution made from the bark and leaves. Headaches were cured by wetting a handkerchief in the solution and tying it around the head. More recently, the same solution has been used to treat saddle sores and swollen legs of horses. Krochmal (1954:9) has attested to the anti-scorbutic qualities of cottonwood.

Mushrooms, a favorite food delicacy of the Cahuilla, were frequently found on dead limbs of the cottonwood (as well as on sycamores, willows, and oaks). Mushrooms found on cottonwoods were referred to as *saqapish,* and those found on the oak were known as *yulal.* The *saqapish* fungi were extremely popular as a food, and when available added a significant amount of protein to the diet. The mushrooms were collected in the spring when they were a whitish color and the sap of the cottonwood had begun to run.

They were still considered good to eat—later in the year when they turned brown. Mushrooms were boiled, fried, and used in making gravy. They were also mixed with acorn mush as a special delicacy.

Proboscidea altheaefolia (Benth.) Dcne. (Martyniaceae)

Common Name: *Unicorn-Plant,*   Cahuilla Name: *akawat*
                 Devil's Claw

The unicorn-plant is found in sandy places of the Lower Sonoran zone on the Colorado Desert. Chase (1919:70) recorded the Cahuilla name as *Pok-ow-wit,* which resembles the name given to a species of snake rather than a plant name. He probably heard *akawat* and recorded the word incorrectly.

The seeds of the plant were used as a food. According to Krochmal (1954:27), the seed is highly nutritious. The hook-like thorns of the plant were used as a tool in mending baskets and broken pottery. The shape of the thorns lent themselves to gripping parts of a basket or vessel while it was being mended.

Prosopis juliflora (Sw.) DC. var. **Torreyana** L. Benson (Leguminosae)

Common Name: *Mesquite, Western*   Cahuilla Name: *ily*
                 Honey Mesquite

Next to the oak tree, mesquite and the screwbean, which is another species of the same genus, were the most extensive food producing trees utilized by the Cahuilla. While the oak tree provided the basic staple food for the Cahuilla Indians of the San Gorgonio Pass and the Santa Rosa Mountains, the mesquite and screwbean were the primary food sources for the Cahuilla of the Colorado Desert. This is not to say that mesquite was unavailable to some groups. As with the oak, mesquite trees could be found within the territory of all Cahuilla groups, but in some areas trees were not as plentiful nor were the beans as palatable as those which were abundantly accessible to the desert groups.

Mesquite is commonly found below the 3,000 foot level on sandy alluvial fans and in washes where the roots can penetrate downward into the water table, often at great depths (Brown 1923: 17). Mesquite is common to the alkali sink of the Salton Basin and prominent stands are found at Torres-Martinez Reservation, Fish Springs, Thousand Palms, Palm Springs, and in Chino Canyon. Mesquite extends as far west as Banning in the San Gorgonio Pass and as far east as the western and southern base of the Santa Rosa Mountains. In the Borrego Desert area, it is in greatest abundance around Borrego Springs and at the openings of Coyote and Rockhouse canyons.

Any survey today can give only a slight conception of the past distribution of the plant, since agricultural and urban development and changes in the water table of the Coachella Valley have resulted in destruction of many of the finest mesquite groves. For example, only scattered clumps of mesquite may be found today near Indian Wells (*Kavinish*), which was aboriginially an extensive gathering area for many Cahuilla groups. Potsherds and other remnants of the past testify to a period when the mesquite groves in this location flourished. According to older Cahuilla, mesquite once spread for many miles along the western border of the Santa Rosa Mountains.

Bean and Saubel (1961:237) have suggested that the stability of many Cahuilla settlements can be explained to a large degree by the presence of oak groves in heavy density, which provided a dependable staple crop. The same hypothesis appears also appears applicable to certain desert Cahuilla groups, who because of large, dependable quantities of available food from mesquite groves were encouraged to maintain a non-migratory life style.

Every part of these deciduous trees and shrubs were useful to the Cahuilla: the trunk, leaves, limbs, thorns, roots, bark, sap, and the pod with its nutritious beans. The mesquite supported certain useful parasites, and it attracted many game animals which sought protection in the dense mesquite thickets. If agriculture was aboriginal among the Cahuilla, the presence of mesquite groves may well have been a contributing factor, since they not only encouraged a sedentary village life, but the Cahuilla also engaged in essentially horticultural practices, pruning mesquite trees and breaking and cutting branches regularly to provide easier access to beans.

Mesquite as Food.—Food products were available at three stages in the annual growth of the mesquite tree, extending from April until August. These stages consisted of gathering the blossoms in the spring, the green pods in early summer, and the mature, naturally dried pods in early autumn.

Curtis (1926:24) reported that the blossoms were picked and roasted in a pit of heated stones, after which they were "squeezed into balls ready for eating." These balls were known as *selkulat* (i.e., "blossoms made of"). Prepared blossoms were stored in pottery vessels and cooked as needed in boiling water. The blossoms were also used in making a tea.

The green pods were available for gathering from early to mid-summer, depending upon the locale. In the lower Colorado Desert, the pods were ready in June; at Palm Springs, in July; and near Whitewater, about August. In a given grove, three weeks or more ordinarily elapsed between the time that green pods were sufficiently mature for harvest and the time when dried, fully mature pods could be picked. The green and dried pods were picked by all members of a family. Children were able to crawl among the branches more easily than adults, however, and pick pods from the center of each tree. To facilitate gathering, tunnels or rooms were made in the mesquite by cutting or breaking branches to provide openings for easy access (Patencio 1943:59).

Since the gathering period spanned several weeks, people usually remained at the gathering site for the duration of the pod-gathering season if groves were any distance from their village site. If the groves were nearby, the people gathered the green beans, carried them home, and returned several weeks later for the naturally dried pods. Even so, there was always the risk of losing a large portion of the crop. When pods began to drop naturally, they might be rapidly depleted as a food source by animals that relished the mesquite bean. Occasionally, other Indians might also poach on a mesquite grove in the absence of its owners. Finally, there was always the possibility of summer thundershowers, when heavy rains caused flash floods that could completely destroy a crop.

The green pods were either prepared at the time of picking or ripened artificially by placing them in the sun. Preparation consisted of pounding or crushing the bean pods in mortars (paal) to produce a pulpy juice. The mortars employed were made of mesquite or cottonwood stumps, two to three feet high. The mortar was prepared by firing the center and then cutting a depression with a stone axe. A wooden or stone pestle, approximately two to three feet long, was used in grinding, enabling the work to be done in a standing position.

Barrows (1900:73) wrote of the pulpy extract produced as follows: ". . . during the hot summer months it is drunk continually. A wide clay basin, containing a mass of half-crushed pods of algaroba [Barrows' word] is kept filled with water, and everybody helps himself to a good draught as thirst impels." According to older Cahuilla, there was always an olla of this beverage around each house. Early reports (Blake 1953:4; Bowers 1888:5; and Barrows 1900:56) all suggest that a light fermentation process enhanced the taste of the beverage. Any suggestion of actual alcoholic content or use of the beverage as a mild intoxicant is denied by present-day Cahuilla.

The mature dried pods could be eaten without any modification immediately after picking. When eaten in this manner, they were often broken up into small pieces about an inch long. Usually the dried pods, however, were ground into a meal in a stone or wooden mortar. Curtis (1926:24) reported that prior to grinding the pods were "parched by stirring them about in a flat dish containing embers." This is the only recorded account of parching, however, and none of the Cahuilla interviewed by the authors can conceive of any reason for such a practice. Possibly, such parching might have facilitated grinding.

The ground mesquite meal was placed in a basket or vessel, dampened with water, and left for a day or so to harden. Barrows (1900:56) said that such meal was referred to as *pechita* or *menyakish*, but present-day Cahuilla agree that *menyakish* refers to mesquite beans at any stage. The hardened meal was sometimes formed into round balls, but more frequently it was molded into cakes ranging in size from two to ten inches in diameter and from one to three inches thick. The larger size was most common. Pieces were broken from these cakes (*kakhat*) and eaten dry, made into mush, or mixed with water to form a beverage. The dried-cake meal was particularly useful to hunters and travelers, since a small amount with the addition of water could provide a substantial meal.

Hooper (1920:357) reported that many of the mesquite beans were "worm-eaten in spots, but regardless of this they are all pounded together." Castetter and Bell (1937:24) said that during storage the ground meal soon became "a living mass, since an insect, a species of *Bruchus* was present in almost every seed." They added: "To the Pima or any other tribe of Indians, this made little difference. The insects were not removed, but accepted as an agreeable ingredient of the flour, subsequently made from the beans. If reduced to a fine flour soon after gathering, the larvae still remained within the beans and became a part of the meal, forming a homogenous mass of animal and vegetable matter."

These statements were contradicted by Bowers (1888:10), who wrote about mesquite stored in granaries as follows: "When filled with pods, they are carefully covered to exclude insects or they will soon be perforated and breed worms." Modern Cahuilla interviewed had various responses concerning the Cahuilla attitude toward larvae-infested beans. Barristo Levi, a desert Cahuilla, said he had always heard that larvae-ridden beans were considered as palatable as those which were not infested. Mrs. Katherine Saubel recalled, however, that when she was a child at Los Coyotes her grandmothers always broke off the "wormy part" of the bean prior to eating. Possibly attitudes toward infested beans varied in different parts of Cahuilla territory or it may have been that new tastes acquired after European contact led to changes in attitudes toward such beans.

Significance of Mesquite.—The primary importance of mesquite as a food source among the Cahuilla was related to several factors aside from its ease of preparation. These factors included the facility with which it lent itself to preservation and storage, its abundance and dependability as a crop, and its rich food content.

Mesquite beans were readily stored for long period of time—up to a year or possibly longer. The meal was stored in ollas or various types of storage baskets, including granaries. Dried cakes were hung from the rafters of homes or—in recent times—wrapped in cloth bags to keep them out of reach of insects and rodents. Cahuilla today recall that the cakes were always available: "The old people always got them, and the children used them for snacks or lunch during the day." Emergency supplies of mesquite beans were often stored in dry caves in higher canyons of the San Jacinto, Santa Rosa, and San Bernardino mountains.

The storage facilities have been well described by Barrows (1900:52) and Bowers (1888:5). Barrows wrote: "In the Cabeson [Coachella Valley] these granaries are made almost exclusively out of the *hang-al*, the species of wormwood so abundant there (*Artemesia ludoviciana* Nutt.), and having been filled with mesquite beans they are covered over and sealed with an armful of the shoots and a daub of mud. These granaries are perched either on platforms of poles, or, in the mountains, on the tops of high boulders, out of the reach of the field mouse or kangaroo rat." Bowers (1888:5) adds several details of his own: ". . . the beans are gathered in autumn and stored in bins for future use. These bins or storehouses are made by twisting willow twigs or arrowweeds into long ropes and sealing one layer over another in a similar manner to the straw rope-bee-hives we see pictured in old books. This is cemented or plastered on the inside and made air tight. They look like huge bulging jars covered with wicker work, and which hold 10 to 15 bushels each. When filled with pods they are carefully covered to exclude insects or they will soon be perforated and breed worms." This quantity of beans was said to be sufficient to feed a family of six to ten people for a year.

Only gross estimates exist of the quantity of beans available to the Cahuilla each year. Barrows (1900:56) reported that a good crop would bend each branch of a mesquite tree almost to the ground and as the fruit fell "pile the ground beneath with a thick carpet of straw-colored pods." Walton (1923:2) apparently was speaking of such crops when he noted that during a favorable season in many areas of the southwest each tree will average one-half to one bushel of beans, the quantities available in an area being "limited only by the facilities available for gathering the fruit." Walton estimated that a single worker could gather about 175 pounds of dried beans in a day, weighing approximately 21 pounds to the bushel or eight and one-half bushels per day. He further estimated that one acre of land well covered with trees could produce 100 bushels per year.

Walton's estimate of gathering possibilities per day must be considered grossly nonselective, however, since the Cahuilla report that the beans of every mesquite tree are not equally palatable as food. In addition, trees varied in producing ability from year to year. Aschmann (1959a:83) pointed out that in central Baja California: "Not every mesquite tree produces significant seed crops in all years, but normally there is some production on some of the trees in any district every year."

Hicks (1963:98) commenting on the reliability of mesquite as a food crop for southern California and Baja California noted that since the growth of mesquite was primarily dependent on ground water, it was not influenced by the vagaries of local rainfall to the same extent as certain other legumes. He recognized, however, that although generally reliable as a source of food, the crop was at times adversely affected. Excessive rainfall or dryness could influence yield in a number of ways. For example, thundershowers coming at a particular time in the ripening process could destroy a crop by causing pods to fall prematurely, the fallen pods rotting in the damp soil or being eaten by animals before they could be gathered. Flash floods also were another danger to mesquite crops, since many of the better mesquite stands were at the mouths of canyons. Bell and Castetter (1937:181) also observed that the mesquite crop was a fairly reliable staple among the Yuman Indians, although "due to rare unfavorable circumstances they too might fail."

Nutritionally, the mesquite beans is said to compare favorably with barley. Bell and Castetter (1937:6) reported that the bean contains a 25 to 30 percent grape sugar content. Data compiled by Foster (1916:4-5) and Garcia (1917:71-82) indicate that mesquite beans per 100 pounds contain 8.34 pounds of crude protein, 52.02 pounds of carbohydrates, and 2.4 pounds of fats.

Other Uses of Mesquite.—In addition to its use as a food, all parts of the mesquite tree—as noted earlier—were utilized for some purpose. The trunk of the mesquite was often used in making wooden mortars. Harrington (unpublished) was told by Adan Castillo that mesquite wood was considered superior to other woods for this purpose. The typical mortar, according to Harrington, was about 30 inches tall and had a hole 15 inches deep. The lower 15 inches of the mortar was buried in the ground. A three-foot pestle was usually employed, sometimes made from a mesquite limb, and grinding was carried out in a standing position.

Smaller limbs of mesquite were considered an excellent material for bowmaking. Hooper (1920:358) described bows of mesquite and desert willow as measuring from three and one-half to four and one-half feet in length and from one and one-half to two and one-half inches in width. Usually the string was fabricated from sinew or agave fiber. Compound arrows were often used that had a fire-hardened foreshaft of mesquite wood inserted in a mainshaft of cane and maintained in place with adhesive mesquite gum.

Mesquite was viewed as one of the best firewoods, it compared favorably with oak and provided a hot, durable fire for cooking, baking pottery, and warmth. Havard (1884:457) said mesquite was unsurpassed as a firewood for such uses. Patencio (1943:123) noted that in pottery-making a pit was formed in which a slow-burning fire of mesquite wood or bark was started. The clay vessel was then baked for one day in the pit.

Mesquite bark was particularly preferred as kindling, for cooking, and as firewood in sweathouses. It was also used as a wrapping. Several years ago, Dr. Clarence E. Smith, then director of Palm Springs Desert Museum, discovered an arrow decorated with small, red, round spots, which was wrapped in mesquite bark. The bark was also pounded, rubbed, and pulled to form a soft fiber used in weaving skirts for women and making diapers for babies. Hooper (1920:367) noted that mesquite bark fiber was used in binding bowls and dishes below their rims to strengthen such vessels. Mrs. Katherine Saubel recalled seeing a carrying net (*ikat*) made of mesquite bark fiber which was used to carry pottery. The net encircled the vessel and could be held like a sack.

Large limbs of mesquite were often used as corner posts for houses, as rafters from which food and household goods could be hung, and as granary posts. Leaves were occasionally used in roofing houses, although palm fronds and arrowweed were preferred.

The mesquite thorn was used in puncturing the skin for tatoos if the preferred cactus thorns were unavailable. Mesquite gum was used in securing foreshafts of arrows, attaching baskets to mortars, and for other adhesive purposes. According to Havard (1888:458) gum is usually found around knotholes of the mesquite, but the yield of gum from a tree could be increased up to as much as a pound of gum per year by making incisions in the bark. The gum was also diluted with water and used as a wash for open wounds, sores, and for treating sore eyes.

Bowers (1888:5) noted that among the desert Cahuilla in the late nineteenth century mesquite beans were used as cattle and horse fodder. In the historical period, mesquite was also used to make agricultural implements. Patencio (1943:79), a Cahuilla from Palm Springs, wrote about such usage as follows:

> After time had passed, the Palm Springs Indians managed to get a few ponies. Then they took the hard mesquite wood, stripped off the bark, and burned the end very hard. This made the plow point. The whole plow was made of one stick or limb. The pony had no harness, only a cinch around its body.
>
> As a little boy I remember how I used to sit upon the pony to guide him, while my father held the plow. The sun would shine nice and warm on my back. I remember how hard I tried to keep awake, but often I fell asleep.
>
> Then for a harrow my father used bunches of mesquite. This, as anyone knows who knows mesquite, makes a very good harrow. There was a good garden made for corn, melons, much garden stuff, and good fields for corn, barley, and wheat. Everything grew, there was plenty to eat at that time.

The shade of mesquite trees was often utilized by women as working areas, where they could grind food out of the direct rays of the sun. The authors observed one such working area in Cottonwood Canyon near Banning, where an awl, broken potsherds, and a buried mortar within a clump of mesquite testified to such usage. Saunders (1914:78), one of the early and more competent desert writers, noted that it was common desert lore that "the shade of the mesquite is the coolest shade in the world." He added: "You see the leaflets are hung in such a way as to admit of their turning with the least air, and that means that the shade is combined with a maximum of circulation."

Mesquite groves served as an indication of ground water supply, and many of the Cahuilla water wells were dug in mesquite groves. Many different kinds of game could also be found in the vicinity of mesquite groves. Large game such as deer, antelope, and mountain sheep ate mesquite pods, and other small game gathered in thse areas for shade, food, and water. Among the most common small game were cottontails, black-tailed jackrabbits, the white-throated wood rat, Gambel's quail, ground squirrels, and various reptiles.

Several insects relished by the Cahuilla were commonly gathered in mesquite groves. The cicada (*tachikal*) was roasted whole on open coals and was a highly favored delicacy. A grasshopper-like insect (unidentified), which the Cahuilla called *pakush*, was captured from mesquite trees and prepared like the cicada.

Two methods of hunting were employed against game in mesquite thickets. Rabbits, rats, squirrels, and other small game were flushed from thickets by poking sticks into them. The men and young boys then shot them with bow and arrow or killed them with a rabbit stick. Another efficient method of securing game was to set fire to mesquite brush as a group enterprise. Children were often used to encircle the mesquite trees, which were then fired. Animals were killed as they attempted to escape or were chased back into the brush, where they burned. Once the fire died down, the burned animals were gathered and prepared for eating. The practice of burning the mesquite also served to thin out stands and resulted in improved crop yields at a later date. Both of the hunting methods mentioned were carried out in the early morning when animals were foraging and the wind was calm, permitting easy control of the fire.

Mesquite and Cahuilla Culture.—The fact that there was ownership of mesquite groves among the Cahuilla reflects the importance of this crop as a basic staple. Each sib or group of related lineages maintained a sib boundary within which separate lineages had specific food-gathering areas protected from encroachment by other lineages or sibs. In a broad sense, a sib owned a large territory, but the actual exploiting or owning unit was the lineage. Usually this meant that one lineage had sole access to a mesquite grove or a designated portion of a grove, such as at Indian Wells where there were a series of large groves and different sibs met to gather mesquite beans.

Several writers have commented upon this ownership of mesquite groves. Strong (1929:47-50) noted that individual clans on the desert had well-defined food-gathering areas, particularly with regard to mesquite groves. These boundaries were well known to the adult men, especially clan leaders, and clans were prepared to fight if any outsider was found poaching in their areas. Hooper (1920:355) also reported that "each village had definite boundaries within which inhabitants lived, hunted, and gathered mesquite and other food products." She added: "Food was very scarce in the old days and any infringement of one group on the land of the adjacent group was considered grounds for enmity and often subsequent war." In regard to the mesquite groves at Indian Wells, Patencio (1943-58-59) wrote: "Thickets of mesquite grew at this place—many of them. The Indian tribes from different places claimed a thicket for themselves, and put their work on it. When the harvest was ripe, all the tribes came and collected their beans. One tribe never took the beans of another tribe." In using the term tribe, it should be noted, Patencio employs customary contemporary Cahuilla usage which corresponds with the anthropological lineage or clan.

Because ownership of such groves was recognized, people could leave food-preparing tools within their own mesquite grove and return to them season after season. Mortars, in particular, which were zealously looked after and privately owned, could be left by women in a grove with the assurance they would still be there when the family returned the next season.

Although the lineage was the major owning unit for a mesquite grove, individual families often owned particular trees which they visited every gathering season. These were the trees that produced heavy crops on a regular basis; less productive or reliable trees were utilized on a first-come-first-served basis.

One method of naming the seasons among the Cahuilla was based upon stages in the development of the mesquite bean. Hooper (1920:363) quoted August Lomas of Martinez Reservation as describing the following eight seasons:

1.	Taspa	budding of trees
2.	Sevwa	blossoming of trees
3.	Heva-wiva	commencing to form beans
4.	Menukis-kwasva	ripening time of beans
5.	Merukis-chaveva	falling of beans
6.	Talpa	midsummer
7.	Uche-wiva	cool days
8.	Tamiva	cold days

In at least two instances, mesquite was used in the naming of Cahuilla lineages. The *Sewaxilyem* lineage of the desert is said to

116

be translated as "mesquite that is not sweet" (Strong 1929:42). The *Sauvel* lineage is said to mean "mesquite that never ripens on one side" or "mesquite that never matured."

Regular trade patterns existed between different Cahuilla groups and between the Cahuilla and neighboring groups such as the Serrano, Luiseño, and Diegueño in which mesquite beans were a significant exchange item. Because of the different food potentialities of each area, trade served to balance food sources. The desert Cahuilla, for example, had less acorns than the mountain-oriented groups, whereas the latter had less access to mesquite. Thus, it was customary to trade surplus foods for other foods or materials. The Wanikik Cahuilla of San Gorgonio Pass, who possessed some mesquite but not as desirable a crop as the desert groups, regularly exchanged their abundant acorns for mesquite beans with the Palm Springs Cahuilla. Older residents at Mesa Grande Reservation (Diegueño) and La Jolla (Luiseño), have recalled that when Cahuilla came to ceremonials or fiestas they brought mesquite cakes as gifts for their hosts.

Since mesquite was an essential food of the Cahuilla, it is not surprising to find that considerable religious attention was devoted to the maintenance of mesquite groves. Religious sanction was necessary before mesquite could be gathered. This particular rite of intensification brought togther all members of a lineage and is referred to by present-day Cahuilla as "feeding the house" (the ceremonial house or *kish'amna'a*). Prior to the ceremony, the *net* (political and ceremonial leader) saw to it that a first crop was picked. This crop was then prepared and eaten in the ceremonial house by members of the lineage group. Afterward, the *net* announced that people could go and gather the mesquite harvest. It was believed that death or illness would be visited upon anyone who gathered food prior to the religious rite.

There were also ceremonies designed to ensure the success of the crop. Shamans, who were able to regulate rain and other natural events through their supernatural powers, used such talents as necessary throughout the seasons. In particular, they used their magic to bring rain in the early spring so that the mesquite crop would be large. When gathering-time came, these same shamans were responsible for preventing rain, since dampness could destroy the crop.

Prosopis pubescens Benth. (Leguminosae)

Common Name: *Screwbean, Tornillo*   Cahuilla Name: *qwinyal*

The screwbean is a less massive tree than the honey mesquite and is lacking in the huge horizontal limbs of the latter. Its leaves are much shorter and its thorns are always smaller. Although screwbean was one of the important food staples of the Cahuilla, especially the desert groups, only casual references to its use are found in literature on the Cahuilla. Like the mesquite, it is found in washes and canyons, usually below 2,500 feet.

After European contact, increased reliance on agriculture by the Cahuilla for their food supply appears to have affected the continuing aboriginal usage of screwbean more than mesquite. Screwbean disappeared as a major staple, except as fodder for cattle and horses, whereas mesquite beans were still heavily used up until a generation or so ago. Today, as is the case with mesquite, most of the aboriginal gathering areas have been destroyed by agriculture or other modern developments, although large screwbean groves still exist at Torres-Martinez Reservation and at Thousand Palms Oasis.

Because screwbean was often found in the same areas as mesquite, both food crops could usually be gathered together, often from adjacent groves. Screwbean pods were picked when ripe and allowed to dry or picked after they were fully dried. Food preparation and usage was the same as with the mesquite bean.

Curiously, Drucker (1937:47) reported that screwbeans had to be artificially ripened in a pit before they were edible, and that several weeks or months were needed for proper ripening. None of the Cahuilla with whom the authors have discussed this statement recall any special treatment for ripening. They all agree that the screwbean could be eaten without any unusual preparation. Pit-curing similar to that described by Drucker has been reported among Yuman groups (Hicks 1961), although what effect it had on palatability of the beans is unclear. Perhaps there were differences in taste values between the Cahuilla and their Yuman neighbors. Two Mojave people told the authors that they ate screwbean in either its natural state or pit-cured, but preferred it pit-cured.

Drucker (1937:47) also stated that screwbean was not utilized by the mountain Cahuilla because it was absent from their gathering areas. His informants, however, were members of the *wi'istam* sib within whose territory the authors have observed screwbean. Surviving members of this group have assured us that screwbean and mesquite were both used as food and gathered in the northern part of Borrego Valley and in Coyote Canyon.

Screwbean had several uses other than as a food. The roots and bark are said to have had medicinal value. The large limbs, like those of the mesquite, were used in construction. Smaller limbs were sometimes used for bows, but it is said they were not ideal material. The mescal cutter, a long pole used to sever agave leaves, was frequently made of screwbean. Information on preservation of screwbean as a food, storage methods, and ownership of screwbean groves is similar to that described for mesquite.

Prunus L. (Rosaceae)

Common Name: *Stone Fruits*   Cahuilla Name: *N.D.*

Three species of *Prunus* with edible fruits are well recalled by the Cahuilla as great delicacies and important foods in late summer and early fall: the desert peach, *P. andersonii* Gray; the western chokecherry, *P. virginiana* L. var. *demissa* (Nutt.) Sarg.; and the holly-leaved cherry or islay, *P. ilicifolia* (Nutt.) Walp. The respective Cahuilla names for these three shrubs and trees were *chawkal*, *atut*, and *chamish*.

Other edible species in Cahuilla territory that were undoubtedly used by the Cahuilla—but possibly less important—were the desert almond, *P. fasiculata* (Torr.) Gray; the desert apricot, *P. fremontii* Wats.; and the bitter cherry, *P. emarginata* (Dougl.) Walp. All of these trees bear edible drupes with a fleshy outer coat and a large pit. They commonly grow in the canyons and foothills in scattered locations throughout Cahuilla territory, some of the species up to 5,000 feet.

Of the desert peach, Barrows (1900:61) wrote: "I found it growing along the eastern summits of the San Jacinto range. Its fruit somewhat resembles the *Zizyphus* and was formerly eaten by the Coahuilla, who called it *cha-wa-kal*." Although no longer gathered today, the desert apricot is remembered as a highly prized food source. In more recent times, the fruit was picked in June, boiled, sweetened by the addition of sugar, and put up in jars as jelly. Mrs. Alice Lopez recalled that the jelly was popular at Santa Rosa, where it was called *tatkaka*.

The western chokecherry commonly grows in the pinyon-pine and pinyon-juniper regions of the San Bernardino, San Jacinto, and Santa Rosa Mountains. About three feet high with wide green leaves, the plant bears fruit which is ready for eating in late summer or early autumn (Sudworth 1908:357). Barrows (1900:61) noted that the western chokecherry was common around "springs and moist canons of the Cahuilla Valley" and bore a "small red berry called *a-tut.*" Older Cahuilla report that the fruit was eaten fresh, but was astringent and could cause stomach trouble. The pit was also ground up and used as a meal. The fruit also could be sun dried for future use. Mrs. Katherine Saubel remembered the plant as growing in great profusion in the vicinity of acorn groves at Nelson's Camp near Los Coyotes.

Cahuilla report that the most keenly favored of the species was the holly-leaved cherry, a dense evergreen shrub or tree, which was abundant in the San Jacinto Mountains. Barrows (1900:61) spoke of the tree as bearing a reddish-yellow fruit resembling a small gage plum, which was "puckery" in taste. According to Barrows, the plums were gathered in quantity in August, spread in the sun until the pulp was dried, and then the pits were broken open and the kernals extracted. The kernals were crushed in the mortar, leached in a sand basket, and boiled into atole.

Modern Cahuilla assert that although some plants occasionally bear bitter fruit, the cherries are usually sweet to the taste. Oscar Clarke has observed that the fruit is usually delicious, and suggests that in all probability Barrows confused the fruit of this tree with that of the western chokecherry described above. Probably both the fruit and the kernals of the holly-leaved cherry were equally popular. Cinciona Lubo recalled that the meal made from grinding the kernals was used to make a tortilla-like food. The fleshy fruit was also pressed to make a drink. Romero (1954:18) reported that the bark was used to make an infusion for curing colds.

The Cahuilla says that at one time more of these trees produced delicious fruit than today, but that a *puul,* angered by his people, caused a bitterness to enter the fruits, and that ever since then the fruit of these trees has been better in some areas than in others.

The gathering season for the *Prunus* species lasted two to three months, depending upon the climate and altitude of trees. If they were not gathered soon after ripening, however, birds destroyed the crop. Hicks (1963:140) noted that trees of the holly-leaved cherry do not "ordinarily occur in groves of sufficient extent to warrant semi-permanent camps for its exploitation." The same may be said of the other species. However, stone fruits were usually within a short distance of acorn gathering sites, and fruit was probably sought out during the acorn harvest. They were also often found within a mile or so of village sites. Customarily, the stone fruits were gathered by women.

Psoralea macrostachya DC. (Leguminosae)

Common Name: *None in Munz*   Cahuilla Name: *N.D.*

This tall herb grows along streams and in other moist places below 5,000 feet. The plant provided a yellow dye in basket-making. The roots of the plant were boiled together with the basket weeds that were to be dyed.

Pteridium aquilinum (L.) Kuhn var. **lanuginosum** (Bong.) Fern. (Pteridaceae)

Common Name: *Brake, Bracken*   Cahuilla Name: *welmat* [?]

Romero (1954:58) reported that this fern provided edible shoots that were asparagus-like and rich in flavor. The shoots were prepared for eating by scraping and then boiling them.

Quercus L. (Facaceae)

Common Name: *Oak*   Cahuilla Name: *None for Genus*

The acorn has long been recognized as one of the outstanding undomesticated food sources of the New World, creating a stable food resource that heavily influenced the socio-cultural development of those societies fortunate enough to possess it. Specifically, cultural evolutionists have observed that the acorn provided a basis for significant social changes. Goldschmidt (1959:190) divided hunting and gathering peoples into two groups—those who were solely hunters and gatherers and those who were basically hunters and gatherers, but who also possessed a significant and attractive food resource and had the technology to exploit it. He referred to this latter stage as sedentary hunting-gathering, adding: "Settled hunters and food gatherers live in areas where the natural supply of food is abundant. In California, the chief source of food was the acorn, an excellent and fairly reliable source of nourishment." Willey and Phillips (1958:134) in their classification of cultural development also recognized that acorn-oriented cultures present problems in classification as compared to other hunting and gathering societies. In California, they noted, there existed a "density of population and relatively advanced level of cultural development made possible by a stabilized food supply in the form of a highly specialized acorn complex." Meighan (1959:305) lent his support to this idea, pointing out that "hunting and gathering cultures in a favored environment may reach equal or greater complexity than some agricultural communities." Others who have commented similarly on the impact of acorn on social development include Beardsley (1956:305), Steward (1955:133-134), and Forde (1952:41-42).

Baumhoff (1963:166) advanced the idea that the economic necessities of acorn transport may "have been largely responsible for determining the minute tribelet areas [in California]." He further suggested that the lack of significant development of aboriginal agriculture in California could be traced to the fact that an agricultural economy in its initial stage would have been less productive than the native economy given an abundant food source such as acorn. Attempts to develop agricultural techniques to the level at which subsistence would have been possible could only have lead to widespread starvation and hardship.

White (1963:116) in his study of the Luiseño (who were neighbors of the Cahuilla) saw several factors associated with acorn gathering and storage as contributing to their relatively sedentary way of life. First, large numbers of harvesters, which were only possible in a sedentary community, were necessary because of the shortness of the harvesting season, the distance to oak groves, and the necessity of gathering large yields that would support the population for most of the year. For this reason, traditional division of labor (men hunting, women gathering) and age distinctions had to be forsaken. Sedentary settlements also made it possible to avoid food loss to animals, since the acorn crop could be protected and harvested before it was lost to birds and rodents. Finally, leaching of acorns encouraged a delimited number of village sites, situated in areas of sufficient water supply for processing of acorns.

All of these views suggest that the acorn complex of the Cahuilla is extremely important to our understanding of their migration and settlement patterns, development of ownership concepts toward the environment, and their attitudes toward division of labor. For many years, there has existed a misconception as to the extent of importance of the acorn in the Cahuilla economy. Barrows (1900:62) wrote that the oak was somewhat rare in the habitat of the Cahuilla and the "acorn is not to them of great economic importance." He added: "They do not put the same dependence upon it as did the Indians along the coast." Nevertheless, the acorn was unquestionably the most significant food source among the Cahuilla, and there was no Cahuilla lineage or village which did not have an association with acorn use. Even people who were young when Barrows collected his data have emphasized to us the importance of acorn in the aboriginal diet.

Certainly, it is true that acorn is not as abundant in Cahuilla territory as in other areas such as Central California, since the bulk of Cahuilla territory is either desert or chaparral country. Oak is abundant above the 3,500 foot level, however, and was generally found within five to twenty miles of most village sites.

Distribution and Species.—Four known species of *Quercus* were used until recently by the Cahuilla: the California black oak, *Q. Kellogii* Newb.; the coast live oak, *Q. agrifolia* Nee; the scrub oak, *Q. dumosa* Nutt.; and the canyon or maul oak, *Q. chrysolepis* Liebm. The Cahuilla names for these species, respectively, were: *qwinyily*, *wi'asily*, *pawish*, and *wi'at*. The late Juan Siva named two other species which were exploited, *i'musily* and *tavasily*, but neither of these have been identified botanically.

The most favored variety was *Q. Kellogii* or *qwinyily*, which was said to have outstanding flavor and the most gelatin-like consistency when cooked, a prerequisite for good acorn mush, which was known as *wiwish*. Marianno Saubel noted that *qwinyily* had to be mixed with other types of acorn mush or they wouldn't congeal properly. This species, which was the only deciduous tree of the four, was said to produce heavy yields with great regularity—far more so than the other species. It also provided more meat per acorn.

The *qwinyily* were found on the high hills, even growing up to the 8,000 foot level in some areas, usually on slopes facing away from the desert. While other oaks were often within a short walking distance of a village, people nevertheless migrated long distances to obtain *qwinyily* acorns. The *wi'at* (*Q. chrysolepis*) was common in canyons and on moist slopes below 6,500 feet, while the *wi'asily* (*Q. agrifolia*) was concentrated in valleys and on lower hills below 3,000 feet. These latter two species were of secondary importance. The third species, the *pawish* (*Q. dumosa*), was common to dry slopes below 5,000 feet. The least favored of the oak, it was mostly used as an additive to other types of acorn meal when there were shortages of the more favored varieties.

The mountain groups of Cahuilla had the advantage of possessing the choicest oak groves. Two of their finest oak grove areas were in the Cahuilla Valley near what is now Anza and throughout the area that is today Los Coyotes Indian Reservation.

The principal oak-gathering areas of the Cahuilla of the San Gorgonio Pass—the *Wanikik*—were the immediate southern slopes of the San Bernardino Mountains to their north and the San Jacinto Mountains to their south, extending as far as Idyllwild. The western edges of the San Jacinto Mountains above Soboba Indian Reservation were also Cahuilla gathering areas.

The Cahuilla groups on the Colorado Desert exploited the least favorable acorn-gathering sites. Acorn was not the most important food resource of the desert groups, although most villages in the Coachella Valley had some groves which were visited annually. They also had economic alliances with groups more favorably located in relation to oak groves. The Cahuilla at Agua Caliente (now Palm Springs), for example, harvested oak trees along the upper reaches of Murray Canyon (about 4,500 feet), and also traded with other groups who had a greater abundance of acorn. Significantly, areas with the most complex social groupings were those with many oak groves, suggesting a correlation between extensive acorn use and socio-political complexity. For example, the maximum number of lineages in a desert area sib were four. In the mountains and San Gorgonio Pass, each sib contained five to ten lineages.

The Cahuilla sense of ownership toward oak groves is indicated by the word for such groves, *meki'i'wah*, which meant "the place that waits for me." Oak groves were owned by lineages, and individual trees within each grove were owned by families within a lineage. More information on ownership patterns in relation to food sources among southern California Indians may be found in Bean (1972) and White (1963).

Crop Yield and Nutrition.—Some oak species produced acorn crops every year, whereas other species produced only every second year. The favorite species, *qwinyily*, bore a crop every second season, but it was considered a regular and good producer, bearing large fruit and averaging between 200 and 300 pounds of acorns in a good season. Although the *wi'asily* bore an annual crop, the acorns were small and the yield was probably less than 100 pounds per tree annually. In addition, Baumhoff (1963:166) pointed out that "the crops are variable, no more than one good one occurring in two years." The crop of the *wi'at* was quite irregular, probably not more than one good crop every three years (Baumhoff 1963:166). Nevertheless, these oak trees yielded 150 to 200 pounds of acorn meat per tree and sometimes up to 450 to 500 pounds. When one realizes that one cup of acorn meal produces about two cups of mush, the significance of such large yields from this food source can be better appreciated.

Baumhoff (1963:162), incorporating the work of other researchers, provided a chemical analysis of acorns for three of the major oak species utilized by the Cahuilla. The results of this analysis are shown in Table I, which includes comparable values for wheat and barley. On the basis of these data, Baumhoff (1963:162-163) pointed out that although acorns were inferior to barley and wheat in protein and carbohydrate content, their higher fat content made them superior to most grains in caloric value, since they average about 2,265 calories per pound as compared with 1,497 for wheat. Baumhoff (1963:163) concluded that generally speaking "acorn compares favorably with the grains in nutritive value."

Although greatly diminished, acorn use by the Cahuilla continues even today. In the fall gathering season, it is not unusual to see people gathering acorns to be taken home, dried, and prepared. More modern methods are now used in preparation, such as a cloth for leaching. Food prepared from acorn meat is considered a delicacy and much favored at social and ceremonial occasions. One of the first things a Cahuilla is likely to ask on arriving at a fiesta or ceremony is whether acorn mush (*wiwish*) is being served.

Acorn Gathering.—The harvest season for acorns was from October to November, occuring just prior to or at the beginning of the first winter rains. If it rained before a crop could be harvested, acorns on the ground became wet and turned black. Although these acorns were said to taste different and were considered unfit for consumption, they were eaten in periods of food shortage. Rainfall coming at the wrong time of year also caused acorns to split and rot on the tree. These acorns could still be gathered, but the damaged blackened areas had to be removed, requiring more work in processing.

When the acorns were ripe for harvesting, hunters or others who had visited the oak groves notified the *net* (lineage leader), who then informed his people. A small gathering of acorns was made and these were then processed either at the oak grove or in the ceremonial house of the lineage. This first crop was then eaten

TABLE I
CHEMICAL COMPOSITION OF HULLED ACORNS

Species	Chemical Composition (in per cent)						
	Water*	Protein	Fats	Fiber	Carbo-hydrates	Ash	Total pro-teins, fats, carbo-hydrates
Lithocarpus densiflora	9.0	2.9	12.1	20.1	54.4	1.4	69
Quercus lobata	9.0	4.9	5.5	9.5	69.0	2.1	79
Q. garryana	9.0	3.9	4.5	12.0	68.9	1.8	77
Q. douglasii	9.0	5.5	8.1	9.8	65.5	2.1	79
Q. chrysolepis	9.0	4.1	8.7	12.7	63.5	2.0	76
Q. agrifolia	9.0	6.3	16.8	11.6	54.6	1.8	78
Q. kelloggii	9.0	4.6	18.0	11.4	55.5	1.6	78
Barley	10.1	8.7	1.9	5.7	71.0	2.6	82
Wheat	12.5	12.3	1.8	2.3	69.4	1.7	84

SOURCE: Baumhoff (1936:162) as modified from Wolf (1945, Table 1) and Spencer (1956, Table 156.).

*Water content of acorns was assumed to be 9 percent on the basis of other analysis (Wolf, 1945, Table 2).

ceremonially by members of the lineage group. It was understood that sickness or death would be visited on anyone who gathered acorns before this ceremony.

During the harvest, most of the men, women, and children in a village moved to the oak groves. This trip took anywhere from one to two days to make, traveling along trails that directly connected the grove to the village. Customarily, the gatherers remained for three to four weeks camped in the oak groves to permit the acorns to dry. If rain occurred during the harvest, it became difficult to dry the acorns. Sometimes, therefore, the trip was shortened by bringing acorns back to the village for husking and drying.

The men climbed the oak trees and knocked acorns to the ground, while the women and children gathered up these acorns and those which had fallen naturally. The husks of the acorns were cracked by placing each acorn on a flat rock with a small indentation and striking it with a smaller rock. The acorns were then laid out for drying. During the weeks in which the acorns were being gathered, dried, and processed, the men also hunted deer and an abundance of small game (squirrels, woodrats, and quail) that could usually be found in the vicinity of oak groves.

Acorn Processing.—Fresh acorns have a bitter, astringent taste, caused by the presence of tannic acid. All acorns therefore required leaching before they could be used as food. Occasionally, children husked dried acorns and ate the nut without leaching. The *wi'at* acorns were said to have the least unpleasant taste when eaten fresh. According to Cahuilla oral literature, acorns were not always bitter. At one time, it is said two powerful shamans had an argument. One of them caused all acorns to become bitter to harm the other shaman. Since that time it has been necessary for Cahuilla to leach their acorns. Another version of the story says that it was the creator-god Mukat who became angry at his people and turned acorns bitter.

Each Cahuilla woman had her own gathering and processing equipment, some of which she left in the oak grove from year to year. The principle implements were carrying baskets (*saqwaval*); the mortar (*qawvaxal*); the pestle (*paul*); leaching baskets (*pachika'va'al*); sifting baskets (*chipatmal*), used to separate coarse meal from fine ground meal; a spoon (*kumal*); and a small handbroom (*alukat*), which was used to clean mortars after grinding. The broom was made of yucca fiber, soap plant, or an unidentified grass. In addition, handfuls of grass were used in cleaning bitter-tasting fibrous material which adhered to the acorn nuts. This material could either be wiped off with the grass or rubbed off with the fingers after wetting of the nuts. The men also had one standard piece of equipment, long sticks used in knocking acorns from the trees. Since most of these pieces of equipment are non-preservable artifacts, few traces of the acorn complex remain at archaeological sites.

Two types of mortars were used to grind the acorns; portable mortars (sometimes used with a hopper) and bedrock mortars. Each woman had her own portable or bedrock mortars. Sometimes she had as many as three or four portable mortars (*qawvaxal*), which were used only by herself or her daughters. Any woman who was careless in her use of mortars was considered a "sloppy" housewife. After use, mortars were carefully cleaned with the handbroom mentioned earlier. Bedrock mortars were always covered with grass after use so that small animals could not get into them. The pestle was left in the bedrock mortar. Portable mortars were turned upside down after use, and usually the pestle was placed under them. Mortars were made by heating the area to be ground out and chipping it with a sharp rock. The depression was then further ground out with a pestle. When the depression was still slight, a basket hopper was attached around it with tar or pitch so that meal could be ground without any loss. Manos and metates (*malal*) were also sometimes used in grinding acorn meal. The significance attached to ownership of mortars was such that after a woman died her mortar was broken and buried upside down.

After acorns were ground, the meal was then placed in a loosely woven leaching basket or an indentation in the sand. A layer of grass, leaves, or other fibrous material served as a lining at the bottom of the basket or sand pit to prevent loss of the meal. Warm or cold water was then poured through the meal several times until the tannic acid was removed. Leaching was said to take longer with cold water. Juan Siva also said that once warm water was used, one could not switch to cold water in the leaching process. As the water was poured through the ground meal, the meal was shaken and moved about to facilitate leaching.

Hayes (1929:50) reported a variation on the customary leaching technique among the Cupeño at Warner's Hot Springs. He observed them soaking acorn and plum seed in the hot springs. Romero (1954:56) also recorded an alternate processing method in which acorns were put into fine, hand-woven net bags tied with rawhide and placed into a stream. According to Romero, the running water caused acorn shells to split open and released most of their tannic acid. They were then dried and ground into meal.

Two grades of ground acorn meal were prepared. A fine meal was used in making acorn bread (*sawish*), which was a meal cake baked in hot coals for several hours. With the introduction of the Mexican tortilla, acorn bread was discontinued among the Cahuilla in favor of the tortilla, which now bears the name *sawish*. Coarse meal was used in making acorn mush (*wiwish*). This meal swells to twice its amount during cooking and turns a pale pink or tan color.

J. P. Harrington (unpublished) was told by Adan Castillo that the mush had to be constantly stirred in cooking to prevent scorching. It was stirred with a palm frond stem paddle, which usually measured about two feet in length and was about 3 inches wide at the bottom and 1½ inches wide at the top. The thickness of the mush was periodically determined by pulling the paddle out of the cooking bowl and letting the mush drip from the paddle. Water was continually added until the right consistency was reached. The cooking process was a delicate one since mush that failed to attain the proper consistency hardened when cold.

128

The skill, sophistication, and care exhibited by a woman in grinding, leaching, preparing, and cooking acorn served to enhance her status. A meal that was reddish and bitter marked its maker as a poor cook. A skilled cook knew how to vary the taste of acorn meal by mixing in different proportions of any of the four types of acorns. She might also add her own special flavorings, such as wheat, chia seed, berries, or meat. Acorn mush jells into a firm consistency like a custard when properly cooked. It can be cut into squares and eaten or eaten directly from a vessel.

Those acorns which were not ground and processed in the oak groves were carried back to the village and stored in granaries. Because these cylindrical granaries were woven from pliable plant materials, air freely circulated through the acorns. The granaries were placed in trees or on posts to prevent rodents from getting into them. Acorns could be stored for a year or more in such granaries. Acorn meal was also preserved by forming it into cakes and allowing them to dry. The cakes could be stored for a long period to regrinding and cooking. For temporary storage and preservation, meal was placed in covered ollas or earthenware pots.

A narrow-necked olla (*tevinyimal*) was customarily used in storing ground meal, which had a flat rock as a cover to keep out pests. Such pottery was made from red clay (*tesnat yulish*), and it was sometimes decorated with red or black paint in a linear or curvilinear design. These storage jars, filled with acorns or acorn meal, were hidden in dry caves or clefts of rock to ensure an emergency food supply. Acorn meal was cooked in another type of bowl (*kavá'mal*).

Other Uses of Oak and Acorns.—Ashes of burnt oak wood and oak bark served a number of medicinal functions. Various healing solutions were made by placing oak ashes in water to produce an antiseptic wash or by soaking bark in water. A large fungus (oak gall), said to have been found frequently on the *pawish* tree, was ground into powder and used as an eye wash or physic.

Romero (1954:55) reported that the bark was used in producing various fast color dyes of a nonfading nature and that a substance made from the bark acted as a preservative in tanning buckskin. He also noted that fallen leaves from oak trees made excellent and warm mattress bedding.

J. P. Harrington (unpublished) quoted Adan Castillo as saying that next to mesquite oak provided the best material for wooden mortars. Castillo also drew a sketch for Harrington of a musical instrument made from acorns. A number of acorns were gathered on a cord, which was then swung against the teeth to produce music. Unhusked acorns were also dried and strung as necklaces. Acorns were used as bait in trigger traps used for capturing small animals. Acorns were also used by children in a game somewhat like jacks and in juggling (one of the favorite entertainment skills of women). Finally, the dried wood of the oak tree was said to produce a hot, long-lasting fire. It was considered an ideal firewood both in producing warmth and for cooking.

Acorns were often used in trade activity that accompanied ceremonial occasions. The Wanikik Cahuilla of San Gorgonio Pass, for example, regularly traded with the Serrano people to the north and the desert Cahuilla groups. Acorn meal was exchanged frequently for pinyon nuts with the Serrano and for mesquite beans and palm tree fruit with the desert Cahuilla. Acorn meal was also used as a payment for special services. If a Cahuilla was treated for an illness by a shaman, he might pay him with acorn meal.

Because it was the most important food staple of the Cahuilla, a bad acorn crop was a serious matter of concern. In some years, it was said that cold weather came too early and trees failed to produce well. Dry seasons also were responsible for sparce crop yields. In particular, the timing of rainfall was said to be critical to a good crop. With the proper amount and timing of rainfall, acorns would double in size.

To ensure good crops of acorns, the shamans (puvulam) and the formal ceremonialists (net and paxáa') carried out rain-producing and rain-preventing rituals. Rain-preventing rituals were always carried out just prior to harvest so that the crop would not be damaged.

White (1963:127) asserted that among the Luiseño Indians acorn crop failures often led to war when large populations found it impossible to obtain acorn surpluses from neighboring rancherias. In such situations, he suggested "warfare becomes an agency of ecology." In the case of the Cahuilla, however, warfare was not a regular mechanism of ecology. Instead, complex economic exchange systems operated through marriage rules and ritual obligations that served as agencies of ecology in most instances (Bean 1972).

When trees failed to produce, the customary mechanism at the family level was for the leader of the family to ask a close relative or neighbor whose trees had produced a good crop if he could gather in his trees. If refused, he often called upon the lineage leader (net), who discussed the problem with other family heads and asked permission for him to gather for that season from the trees of a family with an abundant acorn crop. Similar mechanisms for adjusting a scarcity or abundance of a crop existed at the lineage and sib group levels. Occasionally, however, these mechanisms did break down and warfare did result. More mechanisms for coping with food stress are discussed in greater detail by Bean (1972).

Rhamnus L. (Rhamnaceae)

Common Name: *Buckthorn, Cascara*   Cahuilla Name: *N.D.*

At least two species of *Rhamnus* were used by the Cahuilla. The coffee berry, *R. californica* Esch. ssp. *occidentalis* (Howell) C. B. Wolf, provided berries that were steeped in water and drunk as a laxative or tonic. The buckthorn, *R. crocea* Nutt., although relatively rare in Cahuilla territory, provided an edible berry that was available from August to October.

According to Romero (1954:21), the bark of the coffee berry was stripped off the plant, dried, and ground in a stone mortar. The powder was then used as a cure for constipation. Other Cahuilla interviewed by the authors reported usage of the berries for this purpose as mentioned above. Saunders (1914:141) confirmed a usage similar to that reported by Romero for the Northern Diegueño.

Alice Lopez recalled that an unidentified species of *Rhamnus* which bloomed in May was boiled into a jelly-like substance and eaten. Romero (1954:21) recorded the name of the coffee berry as *hoon-wet-que-wa*, which probably is better rendered as *hunwet qwa'i'va'a* or "what the bears eat."

It should be emphasized that Hardin and Arena (1969:75) have reported that people in Europe have been poisoned by eating the fruit of *Rhamnus* species.

Rhus L. (Anacardiaceae)

Common Name: *Sumac*   Cahuilla Name: *N.D.*

Four species of *Rhus* are said to have been used by the Cahuilla: the sugar bush, *R. ovata* Wats., which the Cahuilla called *nakwet*; the lemonadeberry, *R. integrifolia* (Nutt.) Benth. & Hook.; basketweed, *R. trilobata* Nutt. ex. T. & G., called *selet*; and poison oak, *R. diversiloba* T. & G.

The sugar bush is found in chaparral areas below 3,000 feet throughout much of Cahuilla territory. Its drupe-like fruit, which was about three-quarters of an inch in diameter, was gathered from June until August. The berries were dried and eaten fresh, ground into a flour for mush, or made into a berry. A tea used to cure coughs and colds was made from the leaves of the shrub. Mrs. Edna Badger reported that a white sap exuded on the fruit was used as a sweetener.

The basketweed also produced an edible berry. The shrub was common to chaparral regions up to 3,500 feet, and the authors have observed it in abundance at Piñon Flats. The small, lemony-tasting berries are red in color and about a quarter-inch in diameter. They were ready for harvesting from May to July. The berries were eaten fresh, ground into a flour that was used in a soup, or soaked in water to make a beverage. According to Romero (1954:63), they were also used medicinally as a restorative for inactive stomaches.

The thin, pliable stems of the bush were used for the woof in Cahuilla baskets. A reddish outer layer on the stems was peeled away, leaving a light straw-colored basket-making material. This material sometimes was dyed black by soaking it for a week or more in a solution made from the stems of *Sambucus caerulea* Raf. Palmer (1878:597-598) reported use of the stems in basketry as follows:

The twigs are soaked in water to soften them, and to loosen the bark, which is scraped off by the females. The twigs are then split, by use of the mouth and both hands. Their baskets are built up by a succession of small rolls of grass stems over which these twigs are firmly and closely bound. A bone awl is used to make holes under the rims of grass for the split twigs. Baskets thus made are very durable, will hold water, and are often used to cook in, hot stones being dropped in from time to time until the food is done.

The lemonadeberry is a rounded shrub from three to fifteen feet in height, commonly found in dry places below 2,500 feet. The fruit is reddish and from one-quarter to one-half inch in diameter. The berries, which are slightly acidic, were soaked in water to make a beverage. The fruit was ripe for gathering from June to September.

According to Romero (1954:63), the roots of poison oak were gathered, cut, dried, and used to make a tea. Taken in small quantities, the tea was said to render one immune to poisoning by poison oak. Cahuillas interviewed by the author failed to recall any such use. Hardin and Arena (1969:20) stress that poison oak poisoning can be painful and irritating.

Ribes L. (Saxifragaceae)

Common Name: *Currant, Gooseberry*   Cahuilla Name: *N.D.*

Several *Ribes* species, including *R. malvaceum* Sm. var. *viridifolium* Abrams. and *R. montigenum* McClat., are found growing on mountain slopes and hillsides in Cahuilla territory. They provided a regular and abundant seasonal supply of fresh berries from April through August, depending upon the species and its location.

Ricinus communis L. (Euphorbiaceae)

Common Name: *Castor-Bean*   Cahuilla Name: *navish*

This introduced shrub is common as an escape in waste places. Mrs. Alice Lopez identified it as *navish*, which is also a Cahuilla term that can be applied to any plant considered poisonous.

Castor-bean seed when ripe were crushed into a greasy substance that was said to have medicinal value in relieving sores. Hardin and Arena (1969:91) warn that one to three seeds can be fatal if eaten by a child and can cause severe poisoning in an adult. The leaves and other plant parts are also toxic.

Romneya Coulteri Harv. (Papaveraceae)

Common Name: *Matilija Poppy*   Cahuilla Name: *tewlavel kulux'a*

Cinciona Lubo recalled that a watery substance found in the stalk of this bushy perennial was drunk. The plant is common to dry washes and canyons below 4,000 feet. The Cahuilla name translates as "devil's basket."

Rosa californica Cham. & Schlecht. (Rosaceae)

Common Name: *Wild Rose, Rose*   Cahuilla Name: *ushal*

The wild rose is commonly found in moist places such as cienegas and swamps and along streams in many plant communities in California. Although not abundant in Cahuilla territory, it was frequent enough to serve as a minor food and a valuable source of Vitamin C. The bush is common in the Upper Sonoran and the Transitional Life Zones from 2,000 to 6,000 feet. The bush blooms from May until July, and the buds were picked by the Cahuilla just prior to blossoming and eaten. The blossoms were also soaked in water to make a beverage.

Romero (1954:8) reported that tea made from the blossoms could be used to relieve "clogged stomachs." The beverage made from the blossoms is also said to have calmed infants in pain.

Earle and Jones (1962:229) analyzed the seeds of another species of rose and reported they contained 16.2% protein and 14.2% oils. Zigmond (1941:155-156) considered the rose a nutritious food source, noting that it was higher than most fruits in proteins and has a carbohydrate content equal to or surpassing cherries and blueberries.

Rubus L. (Rosaceae)

Common Name: *Blackberry,*   Cahuilla Name: *pikwlyam*
             Raspberry

Three species of *Rubus* were common to the Upper Sonoran and Transitional Life Zones of Cahuilla territory: the western raspberry, *R. leucodermis* Dougl. ex T. & G.; the thimbleberry, *R. parviflorus* Nutt.; and the California blackberry *R. vitifolius* Cham. & Schlecht. These plants are all small shrubs, usually three to six feet high, and conspicuous for their abundant crops of edible berries.

The authors have seen the thimbleberry in the San Jacinto Mountains and the other two species in thickets in shaded areas of the lower canyons near water. The berries had to be gathered as they ripened or the birds would eat them. Although usually eaten fresh, the berries could be dried and stored for later use. Dried berries were boiled in a small quantity of water. Occasionally, half-ripe berries were soaked in water to make a beverage.

Romero (1954:7) said that the roots of a species he identified as *R. villosus* or wild blackberry were boiled into a tea, which provided permanent relief for mild cases of diarrhea. The species name given by Romero appears to be a printing error. It seems probable that he was referring to *R. vitifolius* or the California blackberry, *R. ursinus* Cham. & Schlecht.

Rumex hymenosepalus Torr. (Polygonaceae)

Common Name: *Canaigre,*   Cahuilla Name: *maalval*
             Wild Rhubarb

The canaigre or wild rhubarb is a perennial common to dry sandy places below 5,000 feet. The authors have observed it in dry washes and on the plains near Banning. Cahuilla report that the stalks were crisp and juicy and were eaten as greens. The roots, which contain up to 35% tannic acid, were used in tanning hides. Hardin and Arena (1969:15) reported that the leaves of the plant are toxic.

Salicornia subterminalis Parish (Chenopoldiaceae)

Common Name: *Glasswort*   Cahuilla Name: *hoat*

The seeds of this annual herb, which is common to low, alkaline places and alkali sinks, were a favorite Cahuilla food. In desert areas, the seeds were available from June to October and were ground into meal. Cahuilla from the mountain areas do not appear to recall the plant. Barrows (1900:57) said it grew abundantly in the Indian Wells area of Coachella Valley, where it sprang up after storms. He recorded the Cahuilla name as *ho-at*.

Salix Gooddingii Ball. (Salicaceae)

Common Name: *Willow, Black Willow*   Cahuilla Name: *avasily*

This willow species is found along streams and in wet places below 2,000 feet in Cahuilla territory. Another species, known to the Cahuilla as *saxat*, was also used by the Cahuilla.

Barrows (1900:49) recorded the use of S. *Goodingii* (synonymous with his S. *nigra*) in bowmaking, although he added that the better bows were made from seasoned limbs of the screwbean (*Prosopis pubescens*). Merrill (1929:239) reported that the willow was used in the warf in basketry. Small branches were used also in weaving large storage and carrying baskets. Mrs. Katherine Saubel recalled that cradle boards were made from willow.

Romero (1954:32-33) reported three other uses for a species he gave as S. *Washingtonia*, which he said was called *ke-cham-ka*. Cahuilla interviewed by the authors do not recall this name, nor does such a species appear in Munz (1965). According to Romero, the leaves were used medicinally. The leaves were ground in a mortar and the pulp steeped in water for several hours. After first taking a bath "to keep the blood at the proper temperature," the patient drank the beverage. In regard to possible medicinal properties, Oscar Clarke noted that salicyclic acid, an active ingredient of aspirin, was first isolated in Europe from a willow species. Romero also reported that the willow was an indicator to thirsty travelers of ground water near the surface.

Salvia L. (Labiatae)

Common Name: *Sage*   Cahuilla Name: *None for Genus*

Four species of sage were used by the Cahuilla: the white sage, *S. apiana* Jeps., known to the Cahuilla as *qas'ily*; the thistle sage, *S. carduacea* Benth., called *palnat*; the famed chia, *S. Columbariae* Benth., known as *pasal*; and the black sage, *S. mellifera*, also known as *qas'ily*. It should be noted that *qas'ily* not only was the name given for two sage species, but appears to have crossed generic lines as a plant name, since it was also used for the saltbush (*Atriplex lentiformis*).

The white sage, distributed widely throughout Cahuilla territory, grows from sea level to 7,000 feet and has the widest range of all of the sage species. Seeds of the plant were gathered from July to September, parched, and ground into a flour for use as mush. Many small seeds used in mush by the Cahuilla were flavorless, whereas sage seeds gave a distinct flavor to mush and were often blended with other seeds. Sometimes the leaves of the plant were crumbled and also added as a flavoring. Earle and Jones (1962:244) reported that the seeds of white sage contain 3.8% ash, 7.9% protein, and 6.9% oil.

As a cure for colds, the leaves were eaten, smoked, and used in the sweathouse. The seeds were used as eye cleansers. One seed was dropped into the eye and allowed to roll around, effectively cleaning the eye. Leaves, crushed and mixed with water, were used as a hair shampoo and dye and as a hair straightener. To eliminate body odors, fresh leaves were crushed and made into a poultice, which was placed beneath the armpits before retiring. This was said to cleanse the sweat glands. Men preparing to go on a hunt often used this technique so that game would not detect human odor.

Leaves of the white sage were also used to prevent bad luck if a menstruating woman accidentally touched a man's hunting equipment. One story told to the authors about this practice occurred in more recent times when a menstruating woman handled her husband's rifle. Subsequently, he was unable to kill deer with the gun. A *puul* (shaman) fixed the rifle by passing it through smoke from the burning leaves of white sage. The rifle dripped "black blood," and the rifle was cleaned. After purification with the smoke, it was said that "the deer would no longer catch the scent of contamination when the man went out to hunt."

The lavender-flowered thistle sage grows in abundance in the Upper Sonoran life zones below 4,500 feet and in sandy, loose soils at about 3,700 feet. The authors have observed it growing profusely over many acres near Anza. The seeds were available for gathering from June to November. They were gathered in great quantities, parched, ground into flour, and mixed with other plant seeds for mush.

The well-known chia plant, whose seeds are sold today in many health food stores, may be found throughout Cahuilla territory below the 6,000 foot level on dry, open plains and in extensive stands in the chaparral areas. In the past, it often covered many acres and was usually readily available near most Cahuilla villages. One of the many instances of plant management practiced by the Cahuilla was the burning over of chia stands periodically to facilitate the next season's growth.

Chia seeds were harvested from June to September by women using a seedbeater. Stalks were bent and the seeds were beaten into a basket. Reportedly, a woman could gather several quarts of the tiny seeds in a few hours. The seeds were hulled in several ways: (1) by rolling them in a metate and applying pressure with a mano; (2) by placing them on a hard surface and walking on them; and (3) by the use of flails in more recent times. Romero (1954:54) also described a more recent practice in which men gathered, cut, and bundled the dry stalks. The stalks were then walked upon to hull them, and chaff was released to the winds. After hulling, the seeds were winnowed in baskets. They were then parched either in the baskets or in clay trays with hot coals and pebbles.

The parched seeds were ground into meal from which cakes or mush could be made. Unground seeds were stored for future use in ollas. Sometimes a beverage was made of unground seeds by soaking them in water. Balls (1965:25) reported that the nutritional value of chia seed is such that one teaspoon was sufficient to keep an individual going on a forced march for 24 hours. Earle and Jones (1962:244) reported chia seeds contain 20.2% protein, 34.4% oil, and 5.6% ash.

Medicinally, chia mush was used as a poultice on infections. The mush was wrapped while hot in a cloth or some other material and applied to the infected area. The seed was also used to cleanse the eyes or remove foreign matter causing irritation in the manner described previously for white sage. It was said that alkaline water could be rendered palatable by adding chia seed.

The black sage provided the Cahuilla with both a food and condiment. This plant was found throughout Cahuilla territory at altitudes ranging from 200 to 2,000 feet. The plant blooms later than chia, extending the gathering period for this species. The seeds were gathered, parched, and ground into a meal. The seeds are highly nutritious and have a rich nutty flavor. Leaves and stalks were gathered in the spring from April through May and used as a food flavoring.

Sambucus mexicana Presl. (Caprifoliaceae)

Common Name: *Elderberry*   Cahuilla Name: *hunqwat*

The elderberry grows throughout Cahuilla territory in canyons and on open flats below 4,500 feet near permanent springs, along streams, and in other moist areas. Barrows (1900:63) reported that the berries were gathered in large quantities by the Cahuilla from July through August. He wrote as follows: "The little clusters are usually dried carefully on the drying floor and so preserved in considerable amounts. When wanted they are cooked into a rich sauce that needs no sweetening. They are delicious thus prepared. An Indian family during this season of the year will subsist largely on these messes of 'sauco.'"

Elderberry plants were usually near village sites, and the berries were readily available in quantity. They were not only used to make elderberry sauce, but were also eaten fresh. Dried berries were stored in ollas for use throughout the year. In recent times, Cahuilla have gathered the berries for making jams and jellies.

Elderberry blossoms were brewed into a medicinal tea for use in curing fevers, upset stomachs, colds, and the flu. The tea was also considered beneficial to newborn babies and good for the teeth. The older blossoms were considered best medicinally. Roots were boiled in water and administered for constipation. It should be stressed that roots, stems, and leaves—and to a certain extent the flowers—of *Sambucus* species contain a poisonous alkaloid and a cyanogenic glycoside that can cause nausea, vomiting, and diarrhea (Hardin and Arena 1969:111).

The elderberry provided two sources of basket dyes. The juice of the berry was squeezed and made a purplish or black coloring for dying basket materials. The stem of the berry was used to make a yellow or orange dye. Twigs of the elderberry were used in making whistles.

Satureja Douglasii (Benth.) Briq. (Labiatae)

Common Name: *Yerba Buena*   Cahuilla Name: *N.D.*

Yerba buena was considered an effective medicine for reducing fevers and curing colds. Plant parts were boiled and drank in an infusion. Palmer (1878:65) reported that a quantity of the plant was bound around the head as a headache remedy. The Luiseño Indians, who are neighbors of the Cahuilla, also had similar uses for the yerba buena plant, which they called *huvamel* (Sparkman 1968:21).

Scirpus L. (Cyperaceae)

Common Name: *Bulrush, Tule*   Cahuilla Name: *pa'ul*

A number of species of *Scirpus* may be found along streams and around springs or in other wet areas of Cahuilla territory from the desert to the mountains. Although *pa'ul* was the name most commonly given for the bulrush, Cinciona Lubo said that *pa'ul* was the name applied to a small species of bulrush. There was also a larger species named *pangawalat* and a species used in house construction known as *pangat*.

The white, starchy tuberous roots of bulrushes were ground into sweet tasting flour. Seeds were gathered and eaten raw or ground into mush. Cakes were made of bulrush pollen. The stalks were used for bedding, mats, weaving materials, and roofing. The ceremonial bundle and images for the Cahuilla image-burning ceremony were also made of bulrush. Merrill (1923:236) reported that the stem of the plant was used for basket wrapping. Since various water fowl nested in areas where bulrush grew, the plants were an indicator for hunters of the presence of game.

Simmondsia chinensis (Link) C. K. Schneid. (Rhamnaceae)

Common Name: *Goatnut, Jojoba*   Cahuilla Name: *qawnaxal*

This shrub, which ranges from three to seven feet in height, is found scattered in stands throughout Cahuilla territory in canyons and on foothills below 5,000 feet. The fruit is about three-quarters of an inch long, brown to black in color, and somewhat resembles an acorn. The seeds are eaten fresh or ground into a powder from which a coffee-like drink can be made. The plant ripens irregularly throughout the year, but usually seeds could be gathered from May until July.

The seeds have a high oil content and there has been periodic interest in cultivating goatnut as a source of industrial oils. Currently, Dr. Demetrios Yermanos, associate agronomist, is conducting feasibility studies of the commercial potential of goatnut oil for the Citrus Research Center and Agricultural Experiment Station at the University of California, Riverside.

Sisymbrium Irio L. (Cruciferae)

Common Name: *London-Rocket*   Cahuilla Name: *N.D.*

Alice Lopez recalled that this mustard plant was used for greens. The leaves were gathered when immature and boiled or fried.

Solanum Douglasii Dunal (Solanaceae)

Common Name: *Nightshade*   Cahuilla Name: *ayka'kal*

This perennial herb is found on partly shaded slopes and in canyons mostly below 5,000 feet. The juice of the berry was used medicinally for sore or infected eyes. The juice was squeezed dirctly into the eye or first diluted with water. It was considered a cure for pink eye and a remedy for eye strain. It was also said that as an eye wash the nightshade improved vision in older people. The dark berries may also have been used as a dye. Hardin and Arena (1969:120) warn that all species of *Solanum* should be suspected of being poisonous. Solanine, a glyco-alkaloid, found throughout the plant, is extremely toxic.

Solanum tuberosa L. (Solanaceae)

Common Name: *Irish Potato, Potato*   Cahuilla Name: *N.D.*

Potatoes probably were first introduced into Cahuilla agriculture sometime in the 1860's or later. None of the earlier accounts of the Cahuilla mention this staple. In the late 1890's, Barrows (1900:71) observed potatoes as one of the crops grown by the Cahuilla.

Solidago californica Nutt. (Compositae)

Common Name: *California Goldenrod*   Cahuilla Name: *pa'kily*

The Cahuilla used the California goldenrod in making a hair rinse and as a medicine in feminine hygiene.

Suaeda Forsk. (Chenopodiaceae)

Common Name: *Sea-Blite, Seep-Weed*   Cahuilla Name: *ngayal*

Several species of these herbs are found in Cahuilla territory, commonly in alkali sinks or salt marshes. The seeds were ground into fine flour for mush or cakes, and the leaves were boiled as greens. Chase (1919:150) reported that *Suaeda suffrutescens* (not in Munz, 1965) was used as a hair dye. The leaves were boiled to produce a dye, which was mixed with clay and applied to the hair, where the mixture was left until dry. Merrill (1923:235) lists two *Suaeda* species used in dyeing basket materials: *S. diffusa* (not in Munz, 1965), and *S. suffrutescens* or *suffracatescens* (also not in Munz, 1965). The plants were boiled in water and basket weeds were left in the water until they reached the desired shade of black. According to Alice Lopez, the dye was also used in making decorated palm mats. Hansjakob Seiler in a personal communication to the authors reported that Matthew Roxy, a Cahuilla, told him that the plant was also used in making a soap.

Taraxacum californicum M. & J. (Compositae)

Common Name: *Dandelion*   Cahuilla Name: *N.D.*

Dandelion stems and leaves were gathered and eaten in spring and early summer. Cinciona Lubo stated that there was a Cahuilla name for the plant, but she was unable to recall it.

Trichostema lanatum Benth. (Labiatae)

Common Name: *Wooly Blue-Curls, Romero*   Cahuilla Name: *N.D.*

The leaves and flowers of this shrub were boiled into a tea for relief of stomach ailments. Alice Lopez stated that one cup of the tea was "good for anything wrong in your stomach."

Trifolium L. (Leguminosae)

Common Name: *Clover*   Cahuilla Name: *tre'evula* [Sp.]

Several species of clover may be found in Cahuilla territory ranging in distribution from the Upper Sonoran through the Transitional Life Zones. Leaves and seeds of various species were gathered from February through July, depending upon their location. The seeds were ground for mush and the leaves were eaten as greens, either raw or boiled. Palmer (1878:424) may have been speaking of the Cahuilla when he reported a method of preparing clover which he observed among a southern California tribe. Large stones were heated and a layer of moistened clover was placed between layers of stones. He noted that young onions and common greens were sometimes cooked with the clover. The Cahuilla name for the plant is derived from the Spanish word for clover, *trebol*.

Triticum aestivum L. (Gramineae)

Common Name: *Wheat*   Cahuilla Name: *pachesal* [?]

Wheat first appears to have reached the Yuman Indian tribes of the Colorado River through Father Kino, who reported it unknown among these Indians until he distributed seeds for sowing (Bolton 1919:I, 373). Wheat quickly attained significance as a crop plant among the Yumans, and Sedelmayr (1939:108,110) observed it in 1774 being cultivated along the lower river. There is a remote possibility that wheat reached the Cahuilla ahead of the Spanish advance (see Appendix). Wheat is mentioned in several Cahuilla myths and Patencio (1943:25) lists it as one of the gifts to the people from their creator-god Mukat. He also gives the Cahuilla name for wheat as *"Pach che sal."*

The first mention of wheat growing among the Cahuilla in the historic period appears in the B. D. Wilson report of 1852 (Caughey 1952:38). Wilson reported that the Cahuilla had a "moderate culture of wheat." Lt. Williamson (1856:98) reported that the Cahuilla had a "good store of grain," which may have been wheat, when the Pacific Railway Survey party crossed the Coachella Valley in 1853.

The Cahuilla parched wheat and ground it into flour. They also often mixed it with the flour of wild seeds such as chia to form a staple mush. During the early historic period, they frequently traded their wheat with travelers passing through Cahuilla territory.

Typha latifolia L. (Typhaceae)

Common Name: *Soft-Flag, Cat-Tail*   Cahuilla Name: *ku'ut*

The soft-flag, a member of the cat-tail family, grows in wet places and marshes below 5,000 feet. The plant was used for food, medicine, and as construction material.

Gathered from June through July, the roots were dried and ground into meal. The pollen, rich in nutrients, was used to make cakes and mush. Zigmond (1941) reported that the pollen compared with rice and corn flours in protein content, but contained less fat.

The roots were also used medicinally to heal bleeding wounds. The stalks were employed as matting materials, bedding, and in constructing ceremonial bundles.

Umbellularia californica (H. & A.) Nutt. (Lauraceae)

Common Name: *Mountain Laurel,*   Cahuilla Name: *N.D.*
        Balm of Heaven, Spice Bush

When the leaves of this aromatic evergreen tree are rubbed in the hand and inhaled or rubbed against the face for a short time, a very distressing headache is produced. The leaves of the plant are also said to have a reverse effect, curing those with headaches. Palmer (1878:652) first noted this curious property, stating: "The Indians of California were aware of the power which this plant had to produce a headache in those that are well and cure those who are affected with it." The tree has a strong, spicy odor, and the Spanish are said to have used the dried, pulverized leaves as a condiment. Although the mountain laurel may attain considerable height, Williamson (1856:364) noted in 1853 that such trees around the Cahuilla village of Martinez were usually from 10 to 20 feet only in height.

Urtica holosericea Nutt. (Urticaceae)

Common Name: *Nettle*   Cahuilla Name: *chikishlyam*

Nettles were used for food, making fiber, in basketry, and as a medicine. The leaves of the plant were eaten raw or boiled as greens. Nettle fiber was used to make bowstrings and processed into cordage. Palmer (1878:649) observed that the fiber is as tough and durable as hemp. The fibers were also used in basketmaking.

Nettles are best remembered among the Cahuilla today for their medicinal value in curing rheumatism and muscular stiffness. Cinciona Lubo recalled that when someone had "stiff feet," nettles were placed on the feet and wrapped in cloth. "It would sting so hard you couldn't help but move," she said. Calistro Tortes recalled his own experience with nettles: "My brothers came to visit me once and my back was sore; they grabbed me and pressed it on my back where it ached. It worked. It was used for all sorts of things like rheumatism—on the arms and back of legs." Juan Siva told the authors he used nettles similarly and also wrapped them around the head as a cure for headaches.

Occasionally, nettles were used to whip children, a rare event since Cahuilla children were rarely punished physically. The stinging caused by the nettles (like numerous insect bites) was very irritating. Cinciona Lubo recalled a child who refused to get up and walk when his mother asked him to do so. His mother placed nettles in the seat of his pants. "He sure moved then," she told the authors.

Vitis Girdiana Munson (Vitaceae)

Common Name: *Wild Grape*   Cahuilla Name: *sawánawet*

The wild grape was common along streams and in canyon bottoms below the 4,000 foot level in Cahuilla territory, often in the vicinity of Cahuilla villages. Grapes were available for picking from June until August and were gathered by women and children. They were eaten fresh, cooked in stews, and dried as raisins. The dried raisins were boiled in water before eating. Cinciona Lubo recalled that in the late nineteenth century a wild grape wine was made by the Cahuilla at Anza. Occasionally, a mush was made from wild grapes. Earle and Jones (1962:239) analyzed samples of two *Vitis* species and found they contained about 10% protein and 16% oils.

Vitis vinifera L. (Vitaceae)

Common Name: *Cultivated Grape*   Cahuilla Name: *N.D.*

The cultivated grape, a European introduction brought to California by the Spanish from Baja California, was grown by the Cahuilla early in the historic period. B. D. Wilson reported in 1852 that the Cahuilla had "producing vineyards" at Agua Caliente, Los Coyotes, and other places, and wrote as if they had been in existence from the mission period (Caughey 1952:37). Wilson assumed that the plantings had been made originally under the tutelage of the missions, but since the Cahuilla resisted missionization fiercely, it seems more likely that they initiated their own vineyards.

Washingtonia filifera (Lindl.) Wendl. (Palmae)

Common Name: *Fan Palm,*   Cahuilla Name: *maul*
              California Fan Palm

The fan palm is common in moist alkaline spots around seeps, springs, and streams below the 3,500 foot level on the western and northern edges of the Colorado Desert. The largest and most extensive native stands in existence today are in Palm Canyon near Palm Springs and at Thousand Palms Canyon near Indio (Benson and Darrow 1944:60). The frequency and nobility of the trees is suggested by the many historic place names alluding to their presence: Palm Springs, Dos Palmas, Thousand Palms, Twentynine Palms, and Two Bunch Palms—to name just a few.

Fan palms range northward as far as Twentynine Palms, first occupied by the Serrano and later by a Chemehuevi group. To the east, fan palms are found extending to Corn Springs in the Chuckawalla Mountains. On the west, their range of distribution is halted by the barriers of the Santa Rosa and San Jacinto mountain ranges. To the south, they extend as far as the Mexican border. Henderson (1951:7-8), who made a comprehensive study of the fan palm, estimates that over 100 groves still exist containing a total of about 11,000 trees. The greatest number of these palms are within the boundaries of Cahuilla territory.

The fan palm was an important economic plant of the Cahuilla, although its usage has been overshadowed in the literature by the more obviously useful plants such as oak, mesquite, and yucca. Nevertheless, it was a regular, dependable, and significant food source to a number of Cahuilla groups and a source of construction and other materials.

Cahuilla oral literature recalls the creation of the first palm tree and its diffusion to other lineage groups in the narrative of the culture hero Sungrey. Patencio (1943:101) tells the story as follows:

One of the head men of Sungrey felt that his time was about gone. His years among his people were many, and he must be prepared to go. This man wanted to be a benefit to his people, so he said: "I am going to be a palm tree. There are no palm trees in the world. My name shall always be *Moul* [sic] (palm tree). From the top of the earth to the end of the earth my name shall be *Moul*. So he stood up very straight and very strong and very powerful, and soon the bark of the tree began to grow around him. And so he passed from the sight of his people.

Now the people were settled all about the country in many places, but they all came to the Indian Well to eat of the fruit of the palm tree. The meat of the fruit was not large, but it was sweet like honey, and was enjoyed by everybody—animals and birds too. The people carried the seed to their homes and palm trees grew from this seed in many places. The palm trees in every place came from this first palm tree, but, like the people who change in customs and language, the palms often were somewhat different, but all, every one of them, came from this first palm tree, the man who wanted to be a benefit to his people.

Use as Food.—The fruit of the fan palm was ready for gathering from late summer until early autumn (late June to early November). Each tree sometimes had as many as a dozen fruit clusters, weighing from five to twenty pounds each. The dark blue fruits are small (about the size of a pea) and have a large seed coated with a thin layer of sweet flesh, which tastes much like the domesticated date.

The fruits were eaten fresh or dried in the sun and then stored in ollas for future use. As needed, they were ground into a flour that included both flesh and seed. The flour was mixed with other flours and water and eaten as a mush. A beverage was also made by soaking the fruits in water. Victoria Wierick recalled also that the fruits were occasionally used to make jelly.

The pith, a spongy tissue occupying the center of the fan palm tree, was sometimes eaten as a famine food (Curtis 1926:178). The pith was called *maul pasun* (heart of the palm) and was boiled for eating. Morton (1963:317-330) noted that the young leaf base of the plant is also edible.

Since fruit clusters usually are out of reach by hand, two methods were used in obtaining fruits. Barrows (1900:36) stated that the fruit clusters were lassoed and pulled from the tree. The authors were told that the more customary method was to detach a cluster from its stem by engaging the stem with a long willow pole with a notch on the end. A twist of the pole disengaged a fruit cluster and it fell to the ground. An innovation in recent years was to attach the lid of a tin can to the top of a pole and slash through the stem.

Other Uses of Fan Palm.—The most common use for fan palm was in house construction. The homes of Cahuilla near palm groves or near enough to trade with lineage groups possessing palm trees were made primarily of the long palm fronds. The fronds were interwoven on the sides of a house and laid over it for a roof. Fronds were regularly changed each year. Houses built of palm fronds were both waterproof and windproof. Ramadas were also made from palm fronds.

146

The seeds of the fan palm were considered excellent filling material for gourd rattles. The palm frond stem was used to make cooking utensils, particularly spoons and stirring implements. Occasionally, bows were made from the stem, but other woods were preferred. The leaves were used for flailing and hulling dried seeds, since they are fibrous and sturdy. In some instances, leaves were used to make *nukily*, the images of the dead that were burned in the Cahuilla memorial rites. Chase (1919:37) recorded the use of palm leaf fibre in basketmaking. If close texture was not essential, loosely-woven baskets were sometimes made from palm leaves. Jane Penn recalled that in her childhood hoops were made from palm leaves and pushed with sticks.

One extremely important use of palm leaves was in making sandals or foot pads (*wakutem*). James (1906:290) observed these sandals still in use in the early part of the twentieth century by older people. The sandals were provided with a loop behind for the heel and two tie-strings in front, one of which passed between the great and second toes and met the other string over the instep.

Palm materials were routinely employed by the Cahuilla in making fire. The origin of the fire-making process is part of the Cahuilla creation myth. Curtis (1926:120) records that in the beginning ". . . they laid a woman, *Ninmaiwaut* (palm), on her back, and *Aawut* (horsefly) took a wooden spindle and drilled her. First blood, then fire, came forth. This woman then became a palm (formerly used for the hearth of the fire-drill), and the man a housefly, which still rubs its sticks together as if using a drill."

Chase (1919:73), an accurate and early observer of the Cahuilla at Palm Springs, recorded the firemaking process as follows:

Two pieces of dry palmfruit stems were the tools, one an inch or so broad, length immaterial, the other less than half as thick, about a foot long, and perfectly straight. A few dead leaves were placed in a little heap; the larger stick was laid beside them and held in place by one of the men, a hollow having first been made in the surface of the wood, with a little groove leading from it to the leaves. Then the smaller stick, trimmed to a blunt point was put to the hollow, and rapidly revolved by rolling between the open hands of the other Indian. His hands moved down as he rolled, returning again and again to the top. The friction sent a fine stream of wood powder down the groove upon the leaves. In less than two minutes smoke showed at the point of friction, then sparks began to fall on the tinder, and finally a flame was started by blowing. Less than three minutes sufficed for the operation. It was hard work while it lasted, for the fire was endangered by the preparation caused on kindling it.

The Palm Oasis.—The larger palm oases, such as Thousand Palms, Palm Canyon, and Andreas Canyon, were favorite habitation sites for the Cahuilla. In areas where there were large palm groves, permanent villages were frequently located. The smaller oases usually were important seasonal camping sites. Such oases on the desert represented ecological niches that offered many advantages, particularly the presence of water, game, edible plants, and an attractive climate.

The palm, an indicator of ground water, usually was found near springs or streams. The stream about the palm oasis in Andreas Canyon, for example, had abundant year-round flow. The spring at Thousand Palms oasis even today provides sufficient water for a small lake. Game is particularly abundant in and near palm oases. Many plants used by the Cahuilla were also found in association with or near palm oases, including sycamore, elder, willow, cottonwood, mesquite, screwbean, arrowweed, yucca, agave, and various cacti. The climate of an oasis is also very conducive to habitation. Such oases are usually on the slopes of hills where desert breezes are frequent and the dense vegetation provides a cool refuge even in the hottest months.

Two important Cahuilla lineages, in particular, were permanently associated with palm oases: the *qawi'siktem*, who lived in Palm Canyon, and the *painiktem*, who lived in Andreas Canyon. The palm groves were owned by lineages and individual palm trees by families. Most other lineages had access to palm oases—even if they were only small stands. Exceptions were the Wanikik Cahuilla of San Gorgonio Pass, who had only a few scattered palms in their area, and the Cahuilla of the San Jacinto area. These groups obtained palm fruit and fronds through trade with other lineages.

The palm oases have for almost a century been favorite collecting areas for the tourist-artifact collector. Most of them are so disturbed today that one can only receive a partial impression of their original state. Although scattered potsherds in palm oases attest to their former occupancy, many artifacts such as mortars and cached ollas have been removed and today are in private collections or museums. Increasingly, air pollution is posing a threat to the existence of the palms, and subdividers are planning to alter such oases as Thousand Palms for commercial developments.

In contrast, the Cahuillas were diligent caretakers of their palm oases—and may even have extended their distribution through plantings. Chase (1919:16) reported that many palm trees growing at higher elevations are known to have been planted by Indians of the area. Harry C. James of Lake Fulmore traced one palm stand in the San Jacinto Mountains to a planting many years ago by a Cahuilla. These examples are all from the historic period. Cahuilla oral literature, however, suggests that palms were planted in aboriginal times, which may explain the heavy distribution of the fan palm in Cahuilla territory. Patencio (1943:101) in recounting tribal history writes: "The people carried seeds to their homes, and palm trees grew from this seed in many places."

The likelihood of such plantings is increased by the fact that the Cahuilla are known to have performed one significant horticultural manipulation in the fan palm stands. Periodically, they set fire to the palms. Several curious interpretations have been placed on this activity. Anthony (1900:237-240) was of the opinion that the trees were burned to send messages to departed friends in the other world. Henderson (1951:7-8) suggested that the trees were burned because they were believed to be hiding places of spirits, an impression that may have been created by noises made by rodents and insects from within the skirts of the fan palm.

Both Chase (1919:37) and Patencio (1943:69), however, state emphatically that the fan palms were burned to improve fruit yield. Patencio writes as follows: "It was the medicine men who burned the palm trees so that they could get good fruit. The bugs that hatched in the top of the palm trees they made the fruit sick, and no fruit came. After the trees were set afire and burned, the bugs were killed and the trees gave good fruit. Now that the medicine men are gone, the worms are taking the flower, the green fruit, and the ripe fruit."

Patencio's explanation of the burning of fan palms as a pest control measure carries considerable weight. A major pest of the fan palm is a bostrychid beetle (*Dinapate wrightii*), the California palm borer (Jaeger 1933:190). Particularly harmful to the fan palm are such other pests as the parlatoria date scale and red spider mites. In the 1930's, the U.S. Department of Agriculture carried out extensive studies of various techniques for eliminating such pests. Although the researchers were unaware of the Cahuilla practice of firing the palms, they recommended periodic firing of the palms as the most effective method for eliminating the scale and red spider mites (Stickney, Barnes, and Simmons: 1950:8).

Yucca L. (Agavaceae)

Common Name: *Spanish Bayonet,* Cahuilla Name: *None for Genus Yucca*

Two species of yucca were abundant in Cahuilla territory: the Spanish bayonet or Our Lord's Candle, *Yucca Whipplei* Torr., called *panu'ul* by the Cahuilla; and the Mohave yucca, *Yucca schidigera* Roezl ex Ortgies, known as *hunuvat.* The Cahuilla were also familiar with two other species which may have occasionally grown near their boundaries: the fleshy-fruited yucca, *Yucca baccata* Torr.; and the Joshua tree, *Yucca brevifolia* Engelm., which was called *humwichawa.* Barrows (1900:59) reported being shown another type of yucca, which he assumed was a distinct species, which was a small plant the Cahuilla called *ku-ku-ul* (*kuku'ul*). A clump of these plants were pointed out to him on the northern slope of Torres Mountain.

Because of their abundance and dependability, the Spanish bayonet and Mohave yucca were important plants. They also provided fiber for many articles of manufacture.

Spanish Bayonet.—The Spanish bayonet was common to dry slopes, particularly chaparral areas, from 1,000 to 4,000 feet. One of the earliest major food plants, this yucca blooms from about April through May in most areas and dies after blossiming. The flower stalk often reaches as high as fifteen feet. The plant was available as a food source for several months to most Cahuilla groups.

Two parts of the plant were used for food: the flower stalk and the blossoms. The stalk was said to be at its best just prior to blossoming when it was full of tasty sap. The stalk was cut near the ground, placed in a rock-lined roasting pit, covered with sand, and cooked overnight. Often after baking, stalks were dried, ground, and mixed with water to form cakes. The stalk was also sometimes sliced, parboiled, and cooked like squash. When blooming had just begun, the blossoms were said to be sweet. As flowers matured, however, they became somewhat bitter. Less mature flowers were parboiled and eaten. Very mature flowers were boiled up to three times with salt before eating. Blossoms could be dried in the sun and preserved. The stalks also could be preserved for a fairly long period. The plant is still occasionally used by Cahuilla today as a food. The flowers now are parboiled and then panfried with tomatoes, onions, and various condiments. The stalk is wrapped in tinfoil and baked in the oven.

Each yucca plant produced food at three stages in its development: (1) when the blossoms were beginning to develop and the stalk was rich in sap; (2) just prior to opening of blossoms; and (3) after blossoming. This permitted a food gathering period of several weeks at any one location and several months if a family collected at different altitudes. Yucca was gathered by both men and women, men usually gathering the plants at some distance from the village. Each plant produced several pounds of stalk and up to five pounds or more of edible blossoms. Although Barrows (1900:50) reported that the fruit pods of the Spanish bayonet were also eaten. Cahuilla interviewed by the authors insist that this was the case only with the pods from Mohave yucca.

Leaves of the Spanish bayonet provided some fiber for various purposes, but the plant was considered inferior to Mohave yucca as a fiber source. J. P. Harrington (unpublished) was told by Adan Castillo that string and rope was never made from Spanish bayonet because the fiber was too stiff and sharp and cut the hands during threading.

The scattered nature of yucca stands did not encourage development of rigid ownership concepts toward specific plant sites. Although a lineage group might claim a particularly abundant site, any member of the group had unrestricted access to the plants. Within sib boundaries, all unclaimed areas were open for gathering to any sib member. This was also true of Mohave yucca.

Mohave Yucca.—The Mohave yucca is a permanent plant with a treelike trunk up to about six feet in height. The plant is found on dry rocky slopes and mesas up to the 5,000 foot level. The yucca is especially abundant from Banning to the Colorado Desert and along the eastern foothills of the Santa Rosa Mountains to the Borrego Valley. The plant occurred within the gathering range of all Cahuilla groups.

As a food source, the Mohave yucca was not as important as the Spanish bayonet. When the fruit pods were green, they could be eaten raw, although they were somewhat puckery to the taste. Usually the fruit pods (*ninyily*) were roasted in hot coals. The fruit had to be gathered early in the season—about April to May— or it became too bitter. The plump green pods are about three to five inches long and an inch or so in diameter. Each pod is filled with four rows of black seeds. Each plant produces several dozen pods and the taste of pods varies in quality on individual plants.

As a soap plant, the Mohave yucca is among the most famous in the Southwest. The roots were scraped and mashed and the shavings were then mixed with water and rubbed into materials during washing.

The plant was most valuable for its fiber, superior to that of most of the other native plants. Large quantities of the yucca fiber were customarily kept about Cahuilla households for a variety of purposes, including use in bowstrings, netting, brushes for body painting, starting material for baskets, and strings for shell money. Krochmal (1954:4) suggests that yucca fiber is equal in strength to any imported commercial fiber used today.

J. P. Harrington (unpublished) reported that the fiber was an ideal material in making ropes and strings, mats, coiled rope soles for sandals, and saddle blankets. In rope making, leaves of the plant from which fiber was to be obtained were soaked until pulpy and the epidermal sheath was gone. The fibers were then buried in mud to whiten them and subsequently combed out. The separated fibers were threaded by rotating a number of fibers on the thigh with the palms of the hand. Spittle or water was used to help bind the fibers in the rope together. It was said that the best fiber was obtained from leaves when they were young and green.

During the Mexican period, saddle blankets woven by southern California Indians, including the Cahuilla, attained widespread popularity and provided a source of income to those who manufactured them. Speaking of southern California Indians generally and the Diegueño in particular, Palmer (1878:646) described the process of making saddle blankets from yucca fiber as follows:

In preparing a warp for the manufacture of saddle blankets, it [the fiber] is first loosely twisted, then when wanted it receives a firmer twist. If the blanket is to be ornamental a part of the warp during the first process is dyed a claret brown, oak bark being used for that purpose. The loom in use among the Indians of today is original with themselves, and not borrowed, as some suppose, from the Spaniards. It is a simply affair consisting of two round, strong, short poles, one suspended and the other fastened to the ground. Upon these is arranged the warp. Two long wooden needles with eyes are threaded with the filling which is more loosely twisted than the warp, in order to give substance and body to the blanket. Each time that the filling is thrust between the threads of the warp by one hand, the Indian female with a long, wide, wooden implement in the other hand, beats it into place. This tool resembles a carving knife but is much larger and longer. One edge is thin, and in this is made a number of teeth or notches not so sharp as to cut.

Barrows (1900:36) reported that the leaves of Mohave yucca were used whole in tying beams and poles in house construction. The seeds were used in making necklaces for women. The seed pod was employed in making toy animals for children; small sticks were added to the pods for legs and eyes were fixed on the front of the pod.

The Joshua Tree.—The Joshua Tree, common to the Mohave Desert, was known to the Cahuilla and may have occasionally grown near the borders of their territory. Juan Siva told the authors that the plant was called *humwichawa* or possibly *hunuvat chiy'a,* since Mrs. Saubel later suggested this was a more likely rendering. He recalled that the fibers were used in making sandals and nets. The Cahuilla also were said to have obtained the blossoms of the plant for food through trade with the Serrano to the north.

The Fleshy-Fruited Yucca.—The fleshy-fruited yucca is common to the eastern part of San Bernardino County and the Joshua Tree wilderness area. The Cahuilla were probably familiar with this plant through the Serrano, but the extent of their usage of it if any is unknown.

Palmer (1871:418) listed the plant as a source of food and fiber among southern California Indians, but did not specify groups using the plant. Both flowers and fruit were eaten and preparation was similar to that reported for Spanish bayonet blossoms and Mohave yucca fruit pods.

Zea maize L. (Maydeae)

Common Name: *Indian Corn,*   Cahuilla Name: *pahavoshlum* [?]
         Maize

Cultivation of corn was aboriginal with a number of Indian groups to the east with whom the Cahuilla came in contact, including the Yumans, Mohave, Pima, Papago, Chemehuevi, and possibly the Kamia. A reexamination of the literature over the past few years and new ethnographic field work has brought together a body of circumstantial evidence suggesting that the growing of corn was probably aboriginal with the Cahuilla (Lawton and Bean, 1968). The problem of aboriginal agriculture among the Cahuilla is dealt with in greater detail in the Appendix.

In 1823, Don José Maria Estudillo, a member of the Romero expedition, which was the first group of white men to cross Coachella Valley, noted the Cahuilla planting corn and three other crop plants, pumpkins, melons, and watermelons, near the vicinity of present-day Thermal (Bean and Mason 1962:46). There are other such reports of corn growing among the Cahuilla during the first half of the nineteenth century.

Patencio (1943:25) gave *"Pa ha vosh lum"* as the Cahuilla name for corn. Mention of corn is made in a number of Cahuilla myths, and an origin account of corn and other crop plants can be found in the Cahuilla creation story.

Corn was ground into meal and eaten boiled by the Cahuilla. During the first month after childbirth, Cahuilla women were expected to subsist on a light diet of corn meal, rice, gravy, and tea (Hooper 1920: 351). Mrs. Alice Lopez reported that corn was sprinkled on the images of the dead during mourning ceremonies in her childhood.

PLANTS
used for
FOOD

shroom *(saqapish)*
cottonwood tree

Pygmy weed
(*Dudleya lanceolata*)

s by Nancy Bercovitz

Nolina (*Nolina Bigelovii*)

Our Lord's Candle *(Yucca Wh*

Photos by Nancy Bercovitz

Fruit of tuna cactus
(*Opuntia megacantha*)

Barrel cactus
(*Echinocactus acanthodes*)

Photos by Nancy Bercovitz

Prickley pear
(*Opuntia occidentalis*)

Beavertail
(*Opuntia basilaris*)

Photos by Nancy Bercovitz

Mesquite (*Prosopis juliflora*)

Cottonwood mortar
for mesquite beans
at Malki Museum.

Photos by Nancy Bercov

Screwbean
(*Prosopis pubescens*)

Photo by Nancy Bercovitz

Mesquite bean granary

C. C. Pierce Photo Courtesy of
Southwest Museum, Los Angeles

Manzanita berries
(*Arctostaphylos glauca*)

White sage seeds
(*Salvia apiana*)

Photos by Nancy Berc

nyon pine
(*Pinus monophylla*)

Pinyon cones: the nuts
have high food value.

Fruit of California fan palm held by Katherine Saubel.
Photo by L. J. Bean

California fan palm *(Washingtonia filifera)*
in Thousand Palm Canyon

J. S. Chase Photo Courtesy of Palm Springs Desert Museum

Juniper berries
(*Juniperus californica*)

Chokecherry (*Prunus virginiana*)

Photos by Nancy Bercovitz

165

Grove of oak trees (*Quercus*)
on Los Coyotes Reservation

Acorns stored in
Cahuilla basket

Photos by Nancy Bercovitz

huilla acorn granary

Cahuilla woman filling granary with acorns

C. C. Pierce Photo Courtesy of Southwest Museum, Los Angeles

Heart of agave is removed
with hardwood pole

Agave (*Agave deserti*

Preparing the heart of agave

Baking agave in the roasting pit

J. S. Chase Photos Courtesy of Palm Springs Desert Museum

PLANTS
used for
MEDICINE

Creosote bush
(*Larrea divari*

Calabazilla (*Cucurbita foetidissima*)

Photos by Nancy Berco

Buckwheat
(*Eriogonum fasciculatum*)

Yerba santa
(*Eriodictyon trichocalyx*)

Croton
(*Croton californicus*)

White sage
(*Salvia apiana*)

Deerhorn cactus (*Opuntia acanthocarpa*)

Tobacco
(*Nicotiana glauca*)

PLANTS

used for

RITUALS

Jimsonweed
(*Datura meteloides*)

Elephant tree (*Bursera microphylla*)

Photos by Nancy Bercovitz

PLANTS
used for
MANUFACTURE

OF BASKETS

Deer-grass *(Muhlenbergia rigens)* bundle
held by Mariano Saubel

Cahuilla woman weaving basket

Photo Courtesy of Southwest Museum, Los Angeles

Katherine Saubel gathers
rush (*Juncus*)

Splitting the reed
in three equal parts

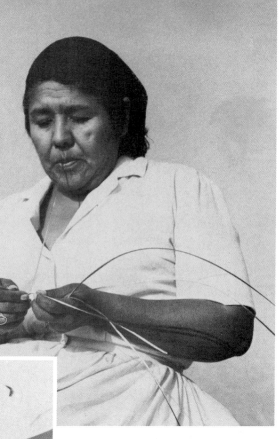

Coiling reed so that
it may be hung to dry

Photos by Nancy Bercovitz

Cahuilla baskets
at Malki Museum

Photos by Nancy Bercovitz

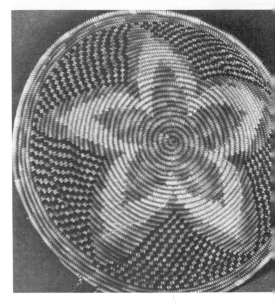

Mohave yucca
(*Yucca schidigera*)

Fibers were used for
sandals, nets, and bows.

Mohave yucca fiber
was used to make
sandals

*Photo Courtesy of
Southwest Museum*

Cahuilla loom used for weaving
yucca fiber into saddle blanket

*Photo Courtesy of Southwest
Museum, Los Angeles*

179

Cahuilla archers lined up. Bows were made of mesquite and desert willow; arrows were made of greasewood and arrowweed.

J. S. Chase Photo Courtesy of
Palm Springs Desert Museum

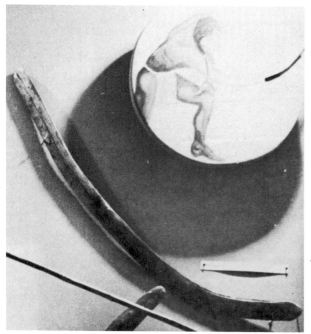

Throwing sticks used to hunt rabbits and ℞
Usually made of ironwood or ribbonwood

Palm Springs Desert Museum

Photo by Nancy Bercovitz

PLANTS
used for
DWELLINGS

Ocotillo
(Fourquieria splendens)

Photo by Nancy Bercovitz

Ocotillo structure

S. Chase Photo Courtesy of Palm Springs Desert Museum

California fan palm
(*Washingtonia filifera*)

Photo by Nancy Bercovitz

Cahuilla home on Colorado Desert

Cahuilla family and dwelling
C. C. Pierce Photo Courtesy of Southwest Museum, Los Angeles

ancisco Patencio and Palm Springs ceremonial house

dwin Photo Courtesy of Southwest Museum, Los Angeles

Reconstructed aboriginal Cahuilla village at Palm Springs Desert Museum.

Photo by Nancy Bercovitz

Cahuilla ceremonial house at Palm Springs

Photo Courtesy of Palm Springs Desert Museum

Cahuilla *kish* (family dwelling) made of tules

Kish of Maria Los Angeles at Cahuilla Reservation

Photo Courtesy of Southwest Museum, Los Angeles

Ceremonial house made of arrowweed (*Pluchea*)
William Duncan Strong Photo Courtesy of Lowie Museum, University of California, Berkeley

Cahuilla Indians threshing in the Cahuilla Valley
Photo Courtesy of Smithsonian Institution, Office of Anthropology

Unidentified Plants

Inevitably, in assembling our ethnobotany, we were left with odds and ends of data presenting unresolvable problems. Particularly frustrating were those Cahuilla names for plants which we could not match up with botanical specimens in the field. In some cases, either one or several Cahuilla were familiar with a plant name in their language, but they had never seen the plant or never seen it in its natural habitat. In other instances, they had seen the plant and were certain they could recognize it in the field. Nevertheless, they were unsuccessful in helping us locate a specimen, possibly because of changes or disruptions in the environment since the plant had last been seen. Sometimes we also encountered disagreement among Cahuilla as to the plant species represented by an Indian name. In addition, there were a number of plants mentioned in the literature which still remain unidentified.

This section presents a list of Cahuilla names for plants which we were unable to identify and the fragmentary information we obtained. Immediately following is similar information on unidentified plants for which no Cahuilla name is recalled. The cacti are excluded here, since unidentified species are discussed under the generic heading. Quite possibly, some of the plants covered in this section are synonymous with plants already discussed. In a few cases, dialectical differences in pronunciation of a name or a failure to recall the Indian name correctly may have prevented identification of a plant in this section with a previously discussed plant. Hopefully, someone may eventually find the information here useful in identifying some of these unknown plants.

Cahuilla Name	Information Available
alnekish	Mrs. Alice Lopez said Ben Levi told her about a plant approximately one foot high found in the Coachella Valley as far west as Indian Wells, which had yellow flowers and edible berries.
alyly	Mr. Juan Siva recalled this as an "ashwood" used in bowmaking.
asil	This plant was said to have a seed resembling that of sage; the seed was ground into mush.
changalangish	Mr. Salvador Lopez, a Cahuilla shaman (*puul*) said this plant was a white-looking little bush with gray leaves. The plant was boiled into a tea for the cure of ulcers.
hunawit	This was identified as a bushlike plant growing in the Santa Rosa Mountains. The green leaves produced a starch substance which many years ago was used to stiffen straw hats.
iculem	Mrs. Victoria Wierick said seeds from this plant were ground up and placed in the roof of the mouth of babies. Water was then given to the baby until it expelled both the water and the seeds, thereby washing out its mouth. Mrs. Alice Lopez was also familiar with the plant name.
isily tuklakma	This plant is said to be "fuzzy." The name means "roof of the mouth of coyote."
iswat	This was described as a small bush, somewhat like a wild lilac. It may possibly be a species of fiddleneck (*Amsinckia* Lehm.)
kawatauna	This plant was used to acquire "power" or luck in the Cahuilla gambling game of peon. Players

Cahuilla Name	Information Available
	rubbed their hands with the plant. According to Matt Calac, it was also used by the Luiseño for the same purpose.
kitavel	Barrows (1900:78) recorded the name of a plant known as *ki-ta-vel*, which was mashed up, roots and all, and used as a poultice or liniment.
kivaal	Cinciona Lubo said this plant was roasted, ground into powder, and mixed with warm water.
kiwut	This plant was described as a bush with a white flower resembling that of another unidentified plant, *wachish*. Fibrous material from the plant was used in making cordage and string for use in house construction. The leaves were used to make brush or a bunch of leaves could be used to remove thorns from cactus.
machil	A type of fern that grows near Idyllwild.
naswet	This was described as a bush which looks like *henil* (*Adenostoma sparsifolium*). Mrs. Alice Lopez believed it might refer to desert lavender or smoketree. Other Cahuilla said it was not a smoketree, but a tree resembling smoketree. Hooper (1920:366) records a myth in which a culture hero, Holinach, lights a branch of *naswit* and throws it into the smokehole of a house.
naxiu	This word is actually a place name in Hathaway Canyon near Banning. It also was said to refer to the sumac.
palqwá'a	A Cahuilla name for an algae or moss said to occur in Millard Canyon near Banning.
palokaiy	This plant was said to grow on Morongo Reservation.

Cahuilla Name	Information Available
palsonul	This plant was said to grow in Coachella Valley. and a Cahuilla geographical site was also named after it.
pachawat	Cinciona Lubo used this name for a sagelike plant with a white stalk, which she said was found in swampy areas near Anza. The green leaves were cooked in shortening (1 teaspoonful) and tied on the forehead to stop nosebleed.
qexawchem	This plant was said to have a small, yellow flower and grow in the hills. The edible seed was parched and ground into flour. The plant may possibly be a species of *Coreopsis*.
qaxáwot	Harrington (unpublished) was told by Adan Castillo that this was a name for kelp, but it is unclear as to whether the name applied generally or to a particular species. Harrington observed that Castillo had a large ball of giant kelp hanging near his house. Castillo said when the kelp became soft it was an indicator of rain. Kelp was sometimes traded from the coast in early times.
saankah	Several Cahuilla used this word in describing an important medicinal plant, which was said to be a sticky, greasy plant, greyish in color, somewhat resembling *Eriodictyon* species. The plant was steeped in water and the solution used to cleanse eyes and wash sores. Leaves were also gathered, ground into powder, and placed on sores to dry them up. Mrs. Alice Lopez, who had used the plant medicinally, said that one first put milkweed on the sore and then put *saankah* powder on it. According to Mrs. Lopez, *saankah* can be used to heal cuts naturally that would otherwise require stitches.
saankat	Barrows (1900:50,77) recorded *sankat* as the plant name for *Adenostoma sparsifolium*, which Cahuilla today identify as *henily*. We were told

Cahuilla Name	Information Available

that *saankat* was the name of a different plant, about ten inches high, seen near Banning. The plant was said to be sticky to the touch (in fact, *saankat* literally means "something that is sticky"). A solution made from the plant was also picked green, boiled, and used as a poultice for bursitis or swellings. The poultice was kept heated during the treatment with a hot rock.

seh'wel George Wharton James (1906:247) reported that the leaves of a plant known as *seh'wel* were rubbed in hot water to produce lather for bathing or washing clothes. This may have been the soap plant (*Chlorogalum pomeridianum*) or some other saponaceous plant species previously discussed.

sevituki Barrows (1900:79) recorded a plant named *se-vi-tu-ki*, which was used to cure fevers. The root was crushed and boiled and the liquid used to bathe the face, neck, and hands. A small amount could also be drunk. None of the Cahuilla interviewed by the authors recall this plant name.

sikwimil This is an unidentified plant said to grow on Section 7 of the Morongo Reservation.

tahat tukal Mrs. Alice Lopez described this plant as a small bush with grayish leaves and small, black seeds enclosed in grayish pods. She said that wherever the plant grows, quail and doves come to eat the seeds. She said the seeds were edible and "tasted good," but she did not remember its particular use as a food plant.

tamapiyikwish The name translates as "something good for the teeth." The plant was described as small with yellow blossoms. It was said to be good for the gums and teeth.

Cahuilla Name	Information Available
tamet mih'a	This plant, whose name means "sun's rays," probably is *Datura meteloides*. It was said to be a poisonous flower that blooms in the morning and "looks at the sun, turning in the afternoon and evening." Some flowers were said to be white and other lavender.
tatuka	This was said to be a bush or tree. The twigs were used in making a tea.
te-a-il	Barrows (1900:79) used this name for a plant whose roots were used in making a medicine for horses. Probably this unidentified plant was *te'ayal* or *Croton californicus*.
tekinat	This was said to be a tall tree. No use of a specific nature was recalled for the plant.
temal	Mrs. Victoria Wierick described a plant which she called "Indian pills" or *temal*. *Temal* is also the Cahuilla word for earth. Cinciona Lubo was familiar with the plant, which she said grew in the mountains and resembled wild poppies. The flower was said to be about two inches in diameter. The stalks were peeled and eaten to quench thirst.
tenil	Barrows (1900:66) reported that an unidentified plant known as *ten-il* was cooked and eaten. He described it as a tall annual with yellow flowers and large leaves. Mrs. Saubel had heard of such a plant and remembered the name. She said that a gum was also made from the plant.
tesail	This unidentified plant is described as having had a yellow blossom used as a dye. The blossoms were boiled and the object to be dyed was soaked in the solution.
tevail	This plant name may have been used for a species of *Croton*.

Cahuilla Name	Information Available

umnawuttuee
This was said to be a starchy plant growing near water or in streams, which has large leaves. It probably is the same plant as *hunawit*, since a substance in the plant was used to starch hats many years ago. It was also said to have a medicinal use.

wachil
Cinciona Lubo recalled a plant about two feet high named *wachil*. She said the green leaves were mashed in water, strained, and placed on the eyes for curing sore eyes.

wachish
This may be the same plant as the one mentioned directly above. The leaves were said to have been used in the same way. They were also said to be useful for making a brush (*kiwal*) and cleaning thorns from tuna cactus. (See also *kiwut.*)

wish
Cinciona Lubo identified *wish* as a bark of a tree which she said was several feet high and hollow. The bark provided fiber for making skirts, cordage, and saddle blankets.

Unidentified Plants Lacking Cahuilla Names

Source of Plant Description	Information Available

Mrs. Alice Lopez

(1) A bushy shrub, referred to as wild lettuce, was said to grow in Saboba Canyon. It was picked in the spring and boiled prior to eating. The taste was said to be spicy.

(2) A saponaceous plant used to wash clothes was rubbed in the water on the object to be washed.

(3) A plant which acted as a coagulant for milk and was used in recent times in making Indian cheese.

(4) The inside bark of a tree was used to make a tea for curing smallpox. The tea was also "very good for you," and Mrs. Lopez said it was used by a Mrs. Alamo of the Torres-Martinez Reservation.

(5) A mintlike plant which grew in well watered ponds and springs. It had a spicy taste and was boiled as a potherb. Oscar Clarke has suggested it may have been a species of *Veronica*. It was also boiled and given to children as a cure for diarrhea and stomachache.

(6) A plant with red berries that were dried, ground into powder, and placed on open sores to facilitate healing.

(7) Two plants growing in the Santa Rosa Mountains were described as being effective medicinally, one for curing toothaches and the other for earaches.

(8) A small white bush was boiled and the decoction drunk to induce vomiting within a few minutes.

(9) A plant that provided a sweet spice and which Mrs. Lopez said was a member of the carrot family.

(10) A plant which grew near Palm Springs and provided a basket dye. The flowers were yellow and tinged with lavender. The flowers were shaken to release the pollen, which was then boiled to make dye.

(11) A mintlike plant that was made into a tea given to women after childbirth to purify their blood. Corn meal was sometimes eaten during the treatment, and acids, fats, and cold foods were

avoided. The plant was used for one or two days after delivery.

Cinciona Lubo

(1) A tall plant about two feet high, growing in meadows, which had yellow flowers. A tea was made from the plant.

(2) A lettuce-like plant which children used as a play cup. The plant was said to be shaped like a cup.

(3) A soap plant resembling sage, but green in color, which grows in the Aguanga area. The plant has a perfume-like odor. It was boiled and used for bathing or washing clothes.

(4) A plant with a root that was boiled and administered as a remedy for stomach pains. Miss Lubo said it would "give you pep and is good for the blood." Two or three dozen of the plants were usually gathered at a time and saved until needed.

(5) A weed which looked like *suul* (see *Muhlanbergia rigens*) and which boys used to catch lizards. The weed was looped on one end, wiggled to attract the attention of lizards, and then used as a lasso to capture them.

(6) A plant with berries that were used as a medicine in feminine hygiene.

Mrs. Katherine Saubel

(1) Our co-author recalled a plant which was used by *ting'ayvachem* (herb doctors) in treating venereal diseases. She described the plant parts used as having burrs that were removed before cooking the parts.

APPENDIX

A Preliminary Reconstruction Of Aboriginal Agricultural Technology Among The Cahuilla

By Harry W. Lawton and Lowell John Bean
Reprinted from *The Indian Historian*, Vol. 1, No. 5, 1968

INTRODUCTION

There is general—although not unanimous—agreement among anthropologists, historians, and geographers of the southwest that all indigenous Southern California Indian groups west of the Colorado River were non-agricultural prior to the founding of Spanish missions along the coast from 1769 to 1821, with the probable exception of the Kamia who may have practiced floodwater farming in Imperial Valley and along the New River as far west as the Salton Sea. Almost invariably, whenever agricultural traits or artifacts have been noted in Southern California, such as the use of corn or the employment of gourd rattles among the Cahuilla of Riverside and San Diego counties, they have been attributed to trade with the

Colorado River tribes. Any case for aboriginal agriculture in Southern California must contend with what has become conventional wisdom. Throughout the literature there is an echo of Father Boscana's (1935:55) statement concerning the Franciscan arrival in Luiseño territory, where ". . . in no part of the province was to be found aught but the common, spontaneous production of the earth."

After establishment of the missions, agriculture is assumed to have filtered inland, finally reaching such unmissionized groups as the Cahuilla, who once they acquired crop seeds from the Spanish and an understanding of the principles of irrigation and other agricultural techniques soon incorporated crop-growing into their culture. In effect, such a conservative view acknowledges rapid acceptance of Mission agricultural technology without taking into account the alternative that similar cultural and ethnic ties should have contributed even more readily to dissemination of agriculture westward from the Colorado River, where floodwater farming was practiced by the Mohave, Maricopa, Cocopa, Yumans, Halchidhoma, Kohuana, and Halyikwamai.

A reexamination of the problems of aboriginal agriculture in Southern California by the authors has indicated there is a strong circumstantial case for its existence among one of the Southern California Indian groups—the Cahuilla. Furthermore, some ethnographic data was found reinforcing the hypothesis of the late Adan E. Treganza (1946) that the Southern Diegueño also possessed agriculture in aboriginal times. The case for aboriginal agriculture among these two groups was recently presented by the authors (Price, 1968). Marginal "kitchen-garden" type agriculture of the sort posited for the Cahuilla and Southern Diegueño may also have existed aboriginally wherever certain environmental conditions permitted among other cultures west of the Colorado River whose subsistence pattern was primarily hunting and gathering. Possible candidates for this "Western Frontier Agricultural Complex" in California are the Cahuilla, Serrano, Chemehuevi, Diegueño, and certain Paiute groups.

This paper is limited to an examination of some of the agricultural practices of the Cahuilla in the historic period. The purpose is not to review the historical and ethnographic evidence suggesting aboriginal agriculture, although some of this material must be cursorily presented for background, but rather to examine the feasibility of aboriginal agriculture. Discussion will center on those agricultural techniques of the historic period that appear to be distinctly native, and which may provide a foundation for eventual reconstruction of pre-Spanish agriculture among the Cahuilla.

CONTACT AGRICULTURE AMONG THE CAHUILLA

At the time of Spanish contact, the Cahuilla inhabited the San Gorgonio Pass, the Coachella Valley eastward to the middle of the present Salton Sea, and the San Jacinto and Santa Rosa mountains and adjacent ranges southward to Warner's Pass. The earliest known recorded contacts between the Cahuilla and Spanish occurred during Juan Bautista de Anza's expeditions of 1774-75 and 1775-76 (Bolton, 1930:I-V). Bolton (1930: I, 100) interpreted the lack of mention of crops in the Anza journals as evidence against Cahuilla agriculture. Both westward passages of Anza through mountain Cahuilla territory by way of Coyote Canyon were made in winter, however, when crops could not have been grown. Both of Anza's return trips to the Colorado River were made in the first week of May, prior to what the authors have found to be the normal season for crops in the mountain valleys. Frost is reported as late as May 6 in some of the mountain areas of San Diego County. Alvino Siva, whose father and grandfather planted crops in a swale at Telyi on Los Coyotes Reservations, informed the authors that frost prevented plantings until about the middle of May.

Thus, even if the mountain groups engaged in agriculture, initial contacts of the Spanish with the Cahuilla occurred at seasons of the year both too late and too early for corn and its associated frost-sensitive crops.

The first Spanish contact with desert Cahuilla groups of the Coachella Valley occurred in 1823 when Captain Don José Romero set out from Mission San Gabriel to establish a new route to the Colorado River. Near present-day Thermal, the Indians were found to have sown corn, pumpkins, melons, and watermelons in the month of December (Bean and Mason, 1962:46).

Almost thirty years elapsed before white men again reported on the Coachella Valley. In November, 1853, the Pacific Railroad Survey party under Lt. R. S. Williamson crossed the Colorado Desert. South of Agua Caliente (now Palm Springs), Williamson (1856:46) observed the "remains of an Indian bush-house and the stubble of a barley field." South of Indian Wells, the surveyors were met by Indians anxious to trade corn, melons, squash, and barley (Williamson, 1856:98). Near Thermal, Williamson (1856:99) reported that the Indians "appear to have a good store of grain and melons, which they raised in the vicinity." East of the Salton Sink, along San Felipe Creek, the party noted fields where crops had been raised (Williamson, 1856:99).

Crops reported by Williamson were also typical of those being raised by the mountain Cahuilla at this late date. The B. D. Wilson report of 1852 (Caughey, 1952:28) stated that the Cahuilla mountain villages had a "moderate culture of wheat, corn, melons, and pumpkins." Strong (1929:38) interviewed elderly informants born in desert villages as far back as the 1850's, and within their memory ". . . both corn and wheat were raised . . . and doubtless other vegetables such as melons, beans, and squash."

None of the literature on the Cahuilla until the late nineteenth century refers to the Cahuilla growing any crop plants other than those associated with the Colorado River agricultural complex. If Spanish agriculture had influenced the Cahuilla significantly, one might expect that they would have adopted certain frost-tolerant vegetables such as onions, cabbages, and beets at an early date. These vegetables—and staples such as potatoes and tomatoes—were grown in mission gardens and raised as early as 1852 in the nearby San Bernardino Valley (Beattie, 1951:202). Although melons, wheat, and barley are European introductions, they were well integrated into the Colorado River agricultural complex prior to Spanish contact with the Cahuilla (Castetter and Bell, 1951:123, 126, 177).

The Cahuilla appear therefore to have remained relatively conservative in their choice of crop plants until a late date. The crop plants they grew were those of the Colorado River tribes. Such was not true of the more westerly and southerly Indian groups of Southern California, who learned agriculture from the Spanish and continued farming after secularization of the missions. Whipple (1858:28), for example, noted that the mission-trained Indians at Santa Isabel rancheria not only grew maize, melons, wheat, and barley, but also raised grapes, peaches, figs, apricots and pears.

FEASIBILITY OF ABORIGINAL AGRICULTURE

The primary argument against feasibility of aboriginal agriculture among the Cahuilla is based on the thesis that the growing of crops in the arid Colorado Desert became possible only after introduction of ditch irrigation by the Spanish. The Colorado River tribes east of the Cahuilla relied mostly on floodwater farming.

On the Colorado Desert, however, there were no native streams for water utilization and a lack of readily procurable surface water (Glendenning, 1951). In the best years, rainfall rarely exceeds seven inches, one inch below the cutoff point for *temporales* agriculture (dry-farming). Average rainfall at Indio is almost exactly three inches. Thus, aboriginal agriculture on the desert might have been helped by occasional rains, but it could never have been dependent upon rainfall. Nevertheless, several geological features compensate for the adverse factors of the environment (Glendenning, 1951:182). Although no streams flow year round, the valley floor is a deep alluvial fill with an underlying aquifer that collects and stores runoff from surrounding mountain ranges. The stratigraphy is such as to create an artesian basin extending from near Indian Wells to the Salton Sea. At various points, faults in impermeable strata occur that during the early historic period allowed water to flow upward in some places, keeping it near the surface. Thus, despite the aridity of the Coachella Valley habitat, it will be shown that neither dry-farming nor conventional ditch irrigation were absolutely essential to crop-growing in the aboriginal period.

Water-Utilization Techniques.—In historic times, at least five natural and artificial water-utilization techniques were employed by the Cahuilla to grow crops. These methods included: (1) dry-farming or *temporales* agriculture, dependent upon rainfall in excess of eight inches annually; (2) conventional irrigation by ditches from wells, springs, streams, and small impoundments; (3) diversion of artesian flow so that it flooded and soaked gardens prior to planting; (4) runoff farming or the exploitation of small rainfall catchment basins with soil-moisture storage capacity; and (5) pot irrigation. Each of these methods is discussed below.

Dry-Farming.—Dry-farming was possible only at mountain Cahuilla villages. Rainfall is relied upon to grow wheat in the Cahuilla myth of "Kunvachmal and Tukvachtaha." where the culture hero also shows himself aware of the fact that too much rain soon after sowing washes away crop seed (Hooper, 1920:368-69). Barrows (1900:71; field notes) reported that wheat and barley were raised in the 1890's in the mountains without irrigation, and that snow and frost often ruined crops.

Conventional Irrigation.—Diversion of water from streams and springs by ditches and the impoundment of water in canyons with small dams was established among the Cahuilla by the mid-nineteenth century (Caughey, 1952:28). Patencio (1943:56) wrote of a hand-dug ditch in Chino Canyon that was used by the "grandfathers and great-grandfathers of the oldest people of our tribe" to divert water to a garden where crops could be cultivated. Other references to diversion ditches and the use of small dams occur in Patencio (1943:57-58) and Barrows (1900:71).

Although such conventional water-regulation methods probably represent a diffusion from the Missions or early Mexican settlers, acquisition of such techniques from Indian groups to the east cannot be totally dismissed. The Yumans sometimes diverted flood-water into swales, the Gila-Pimans had a well-developed canal irrigation system, and the Papago employed canal irrigation from streams aboriginally (Castetter and Bell, 1951:133, 239; Spier 1933:55). Dams were reported on the California side of the Colorado River as early as 1776 (Garcés, 1963:66). Thus, prior to establishment of the missions, there was an opportunity over hundreds of years for the Cahuilla through intermarriage, military alliance with tribes such as the Halchidhoma, and trade with Indian groups to the east to observe and learn about various water-utilization techniques.

Nor can we dismiss the remote chance of independent invention of ditch irrigation and water impoundment by dams. Steward (1929:15, 1938:53) speculated that ditch irrigation of wild seed plots among the Eastern Mono may have occurred to them through the discovery that wild plants grow more prolifically in the swampy low-lands of Owens Valley, although he conceded the possibility of Spanish or American influence. Downs (1964:39-50) showed that environmental manipulation occurred with relative frequency among Great Basin hunting and gathering groups. Even the non-agricultural Washoe diverted streams and ponded water with dams to make fishing easier (Downs, 1966:15). Independent invention of ditches and dams by the Cahuilla might have been an extension of earth-moving skills we know were aboriginal with these people.

Williamson (1856:98) reported that the Cahuilla deepened and enlarged springs. He also observed two springs with raised embankments around them, which he believed the Indians had formed by cleaning out debris and sand. Near Thermal, Williamson (1856:99) found a pool 20 feet or more in diameter at a Cahuilla *rancheria*, which had been "created by an artificial embankment three or four feet high."

Both the Cahuilla and their immediate neighbors to the east, the Kamia, perfected one highly sophisticated water supply technique —the water well. A Cahuilla well reported by Romero (Bean and Mason, 1962:37) in 1823 had a depth of about fifteen feet. These wells were great pits dug into the desert sands with terraced steps making it possible for a woman with an olla to descend to the narrow hole at the bottom and dip up water.

Thus, earth-moving skills acquired by the Cahuilla in cleaning out springs may have been extended to enlarging springs with artificial embankments and finally evolved into digging wells. It would not appear to be such a great technological leap from this point to conducting water short distances out of the mouths of canyons with rude ditches or trapping canyon waters with earthen dams. As Treganza (1946:171) noted, the basic principle underlying canal irrigation is the concept of water diversion.

Artesian Flow.—Glendenning (1951:187) reported that in the late nineteenth century, before white settlers lowered the water table by drilling hundreds of wells in the Coachella Valley, there were many seepy areas in the artesian section. There are reports in the literature indicating that in the early historic period there were marshy areas in parts of the valley created by breakthrough in artesian flow (Bean and Mason, 1962:37; Williamson, 1956:249; Strong, 1929:43-45).

Strong (1929:43-45) found that agricultural plots were cultivated in four Cahuilla villages around 1850, and natural seepage was recalled as the water source for crops at one of these villages. Fields reported by Williamson (1856:105) along San Felipe Creek are also in an area where seepage occurs throughout the year even today (Ruibal, 1959).

The moist soil in the vicinity of springs or places where faults in the impermeable strata had channelled water upward from the underground acquifer could well have provided sufficient moisture in the aboriginal period for growing crops in some desert locations. Gifford (1932) reported an analogous instance of the Southeastern Yavapai growing crops in the moist soil adjacent to a spring. Older Cahuilla, whose grandparents farmed in the Coachella Valley, recall that when they were young there was often sufficient moisture near the surface to plant crops in certain mesquite groves, and that their ancestors sometimes tapped artesian flow that was close to the surface in order to divert the water to agricultural plots prior to planting.

Runoff Farming.—Runoff farming is a method of arid-environment agriculture that is ancient and widespread in agricultural history. The essentials of this system are a relatively large natural watershed area which concentrates its rainfall runoff into a small catchment basin, usually amounting to about an acre or less.

Runoff farming was occasionally employed by Indian groups of the southwest in areas where there were few other means of water utilization for agriculture. It has been reported for both the Papago and Pima (Underhill, 1936:6; Ezell, 1961:9, 148).

Runoff farming among the Cahuilla can be deduced as early as 1853 from the Williamson report, which precisely describes the terrain surrounding a Cahuilla barley field. Williamson's (1856:96) description is as follows:

Several drifts and broad thin layers of blown sand were passed. The accumulations vary in depth from a few inches to fifteen feet; and the surfaces were beautifully smooth and rounded, and generally covered with ripple marks, similar to those produced under water. As we proceeded, we found this sand rising in high drifts, which bounded our vision on the left, while on the right, the base of the mountains was not far distant. A narrow, but nearly level valley was thus formed. The soil appeared to contain a large portion of clay mingled with sand, and several low places, where water had been standing, were covered with a thin coat of fine clay, now cracked and curled up. Near one of the low places, we found the stubble of a barley field. This barley had been raised at the foot of one of the highest sand drifts. The sand was thus found to have a bluish-gray color and to be very compact and hard.

Williamson (1856:249) reported that the thickness of the stubble was such as to indicate a "large yield" of barley. Dr. Walter Reuther of the University of California, Riverside, a specialist on soil and irrigation problems of the Coachella Valley, examined the Williamson data for the authors and termed it an accurate description of topography and soil conditions near Indian Wells. He concluded:

The barley field was evidently situated in a small natural catchment basin that collected runoff water and permitted the growing of a barley crop. This was evidently an area of perhaps an acre or two with deep soil having a high moisture-holding capacity. There are many such small natural catchment basins in the Coachella Valley where soil might remain relatively moist well into summer. In such basins, the effective annual rainfall may be multiplied several times by moisture supplied from watershed runoff after rains.

Runoff farming is also reported among the Cahuilla at *Telyi* on Los Coyotes Reservation, where Alvino Siva's grandfather planted crops in a moist swale in the late nineteenth century. An instance of runoff farming among the Southern Diegueño has also been reported (Florence Shipek, unpublished).

Pot Irrigation.—Mrs. Alice Lopez informed the authors that in the 1890's there was often insufficient water at the village of Toro to grow crops. She and other young girls were sent a considerable distance to bring back water to sprinkle on the vegetables. She said her grandmother stated that in the "old days" ollas would have been used as containers in watering crops.

Among groups to the east of the Cahuilla, pot irrigation has been reported as a sporadic practice among the Mohave, Chemehuevi, and Maricopa; the Pima and Papago sometimes watered gourds by hand (Castetter and Bell, 1942:172; Castetter and Bell, 1951:40, 143, 172).

The development of well-digging among the desert Cahuilla is suggestive of pot irrigation. Both Barrows (1900:71) and Bowers (1885:5) reported that Cahuilla water wells were used for irrigation, although neither specified how it was accomplished. Pot irrigation would certainly represent a logical solution. Decline of the water table in the Coachella Valley in the late nineteenth century may have made it necessary to use wells only for domestic purposes, thus resulting in a loss of specific recollections by informants.

PLANTING, CULTIVATION, AND OTHER AGRICULTURAL TECHNIQUES

The Cahuilla planted crops in small patches (Bean and Mason, 1962:46), which is not suggestive of mission row-crop agriculture, but does correlate with Colorado River practices (Castetter and Bell, 1951:151). Edward Kintano of Indio recalled that his grandparents often planted squash and beans in tiny patches adjacent to or beneath clumps of mesquite and arrowweed (Allen, 1968).

The planting stick was employed (Drucker, 1937:11), and harrowing was accomplished with bunches of mesquite (Patencio, 1943: 79). Agricultural plots on the desert were owned and exploited by individual clans (Strong, 1929:40). Women assisted men in planting—as was customary among the Yumans—but men performed weeding and other heavy tasks associated with crop-raising, such as granary building, while women harvested, prepared, and stored the crops (Bean, field notes).

Plantings among the Cahuilla apparently were made at least twice annually. Estudillo reported crops were sown in December of 1823 near Thermal (Bean and Mason, 1962:46). The Cahuilla must have suffered occasional frost losses in January or March in the Coachella Valley, and at times must have had to replant winter sowings. Planting of crops adjacent to or under mesquite or arrowweed clumps probably was carried out as a frost-protection measure. A late summer or early fall planting is indicated by the fact that Williamson (1856:98) was met in mid-November by Indians seeking to trade their crops, including melons which were not subject to lengthy fresh storage.

Crop seeds were stored in ollas for future plantings. Reed (1963:111) reported finding an olla in the 1930's in a cave at Middle

Willows in Coyote Canyon that was sealed with fibre and rocks and contained watermelon, pumpkin, muskmelon, and sunflower seeds. Treganza (1947:169-170) reported a similar find for the Southern Diegueño, and among the lower Colorado River groups sealed ollas were also used for crop storage (Castetter and Bell, 1951:164).

The Cahuilla followed the practices of the Mohave, Yuma, and Cocopa (Castetter and Bell, 1951:161) in storing both maize and mesquite in large granary baskets of intertwined willow branches. Pumpkins, melons, and squash were cut into strips and dried in the same fashion as among the Colorado River groups for preservation (Castetter and Bell, 1951:112).

The very concept of an agricultural plot on the desert appears to have been viewed in the context of the ecological situation. The late Juan Siva, born in 1875, stated that his grandfather planted crops at the southern edge of Rabbit Peak near Martinez and at Thousand Palms in the Borrego Valley. Siva defined *pawisisual* as "place where you can plant things" (Bean, field notes). The *can* suggests just how difficult it was to find the right planting niche. Moisture was not the only problem, since alkaline soils throughout much of the Coachella Valley precisely delimit the areas capable of cultivation (Glendenning, 1951:83). Drought in some years or shifts in artesian flow may have required the selection of new planting sites. Diminishment of yields through continued cropping also probably forced the Cahuilla to periodically abandon agricultural plots.

In choosing new sites, the Cahuilla may have had a better guide to those perennially moist micro-niches capable of supporting crops than the later Anglo homesteader who could only visually identify areas of obvious alkalinity and whose row-crop form of agriculture would have led him to reject small, fertile areas that would have been favorable to the kitchen-garden farming of a small Cahuilla lineage.

In the Cahuilla creation myth, a blue frog with three stripes down its back is associated with the origin of agriculture as the slayer of the god Mukat from whose body sprouts the first crop plants: tobacco, corn, watermelons, squash, beans, and wheat (Strong, 1929:130-143). Indigenous to the Coachella Valley, there is such a blue frog, *Hyla regilla*, found only where there is year-round moisture. Although the adult frog survives in brackish, saline seepages, salinities above two or three thousand parts per million are lethal to its eggs (Ruibal, 1959:317). Thus, chorusing during mating season occurs only in areas of milder salinity. Since crop tolerances to salinity are lower than *Hyla regilla* eggs, the frog would not be an infallible indicator of agricultural niches, but it could have been a guide to areas in which crops were most likely to grow without irrigation.

As among various Indian groups to the east, star lore appears to have had some association with agriculture. Mr. Mariano Saubel told the authors that observations of the moon were employed to foretell late frosts. His wife, Mrs. Katherine Saubel, said that the position of Orion determined if it was too late for planting. Patencio (1943:113) reported that growing crops was regulated by the "signs of the sun and moon."

PROTOAGRICULTURAL MANIPULATIONS

Downs (1966:54) used the term "protoagricultural manipulations" to describe those incipient agricultural techniques among a hunting and gathering people that are designed to increase the supply of wild plant food, improve bearing, or insure the appearance of wild plants in a particular area. All three of these types of environmental manipulations of wild plants can be found among the Cahuilla and their existence may have contributed to the Cahuilla acceptance of agriculture in the aboriginal period.

Medicine men grew and cultivated their own special plots of medicinal herbs and tobacco, tubers were replanted in areas near villages, and grass was periodically fired to increase the wild seed crop (Bean, field notes). Patencio (1943:91-95, 99-102) speaks of native palms being planted at oases by Cahuilla.

Whether fallowing was employed with agricultural crops is not known, but Patencio (1943:69) reported one type of fallowing associated with native plants. Gathering of fruit was sometimes discontinued for one year to "let the ground have more seeds for plants another year."

Some horticultural practices were performed to assist harvesting or improvement of wild plant crops. Mesquite was pruned by breaking and cutting branches to provide easier access to beans (Patencio, 1943:59). Native palms (*Washingtonia filifera*) were fired by the shaman to kill pests and diseases causing damage to the trees and decreased crops (Patencio, 1943:69). Interestingly, this technique was rediscovered in the 1930's by the U.S. Department of Agriculture as an effective means of killing parlatoria date scale and red spider mites (Stickney, Barnes, and Simmons, 1950:8).

THE LAKE CAHUILLA HYPOTHESIS

Forbes (1963:3) has commented upon the ambiguity of the term "aboriginal" if it is viewed as the possession of cultural traits totally indigenous and free from alien influences. Even the native agriculture of the Colorado River tribes fails to qualify under such a definition, since their agriculture itself was acquired by diffusion as were a number of European crop plants that reached them in advance of Spanish contact. Agriculture made its way to the Cahuilla either via diffusion from the Colorado River complex or it was mission-derived. The authors are of the opinion that Cahuilla agricultural traits recorded in the historic period indicate Indian diffusion rather than Spanish introduction. Diffusion may have occurred relatively early or come about prior to Spanish contact with the Cahuilla but soon after Spanish contact with the Colorado River tribes.

In considering the possibility of an early diffusion to the Southern Diegueño, Treganza (1946) suggested a hypothesis that is even better suited to the possible aboriginal origin of agriculture among the Cahuilla. In pre-contact times, after the formation of Lake Cahuilla in the Salton Sink (Coachella and Imperial Valleys), certain Yuman groups are believed to have moved into the area and settled around the lake shore (Rogers, 1938:122). The fresh-water sea, about 85 miles long and 35 miles wide, existed from about 900 to 1500 A.D. When the Colorado River shifted its course into the Gulf of California, the sea began drying up, probably at about 10 feet per year (Aschmann, 1964:244). The disappearance of Lake Cahuilla compelled those tribes of Yumans on the east to move into the mountains of San Diego County, according to Rogers (1938:122). Treganza (1947-170) speculated that the Yuman groups may have practiced agriculture of the floodwater type along overflow channels leading into the sea. They might more easily, however, have practiced agriculture around the shores of the lake. The lake would have produced a water table which would have supplied surface soil moisture by capillary action in irregular areas along the changing shore. This arable land would have been relatively free of salinity. Treganza suggested that knowledge of agriculture, if not its actual practice, may have been transmitted from Lake Cahuilla westward to the Jacumba area and Southern Diegueño.

None of the literature examining Yuman occupation of Lake Cahuilla has yet come to grips with the problem of the Cahuilla occupants of the western portion of the valley. A number of Cahuilla myths, however, make mention of the lake and the tribal dislocations that occurred among the Cahuilla as a result of its formation (Strong 1929:87). Assuming that the Yumans practiced agriculture along the lake, transmittal of this knowledge a few miles west to the Cahuilla would seem even more likely than its diffusion over intervening mountain barriers to Jacumba. The Cahuilla might well have emulated the Yumans in planting around the shores of the lake, only to discover later as the lake subsided that there were occasional seepage areas, natural catchment basins, and other favorable micro-niches making it possible to plant crops season after season without conventional irrigation.

CONCLUSIONS

We have seen that the crucial factors in maintenance of Cahuilla agriculture in the historic period were not one specific feature such as ditch irrigation, but a number of skills indicating an intimate understanding on the part of these people of their rigorous environment and how it might be exploited for crop-growing.

The authors have established that in the historic period water-utilization techniques for agriculture other than those common to the Spanish existed among the Cahuilla. They have shown that these techniques could be found among the Indian groups to the east of the Cahuilla and that the Cahuilla had an opportunity to learn such methods and adapt them to their own environment.

The failure of the Cahuilla to adopt certain cold-hardy European crops that might have endured the Coachella Valley frost season, the frequent similarity of agricultural traits to those of the Colorado River complex, and the fact that their crops remained typical of that complex until late in the nineteenth century suggest the strong commitment of the Cahuilla to their own agricultural system. The absence of significant Mission agricultural elements in the Cahuilla kitchen-garden type of agriculture and the presence of protoagricultural manipulations for wild plants suggests that their agricultural was aboriginal.

ACKNOWLEDGEMENTS

The authors are indebted to the following persons for information, assistance or advice: Drs. Sylvia Broadbent, Michael Kearney, and Eugene Anderson, all of the Department of Anthropology,

University of California, Riverside (UCR); Dr. John M. Steadman, professor of English, UCR; Dr. Walter Reuther, professor of horticulture, UCR; Dr. Cyrus McKell, professor of agronomy, UCR; Dr. Rodolfo Ruibal, professor of zoology, UCR; Florence Shipek, San Diego; Thaddeus Allen and Laurie Pappas, California State College, Hayward; and the Faculty Research Committee of California State College, Hayward.

We wish to express our deepest gratitude to those Cahuilla people who were interested in our research and helped provide information on agriculture; Mr. and Mrs. Mariano Saubel, Mrs. Jane K. Penn, and Mr. and Mrs. Williams Holmes, all of Morongo Reservation; Alvino Siva of Beaumont; Mrs. Alice Lopez of Banning; and Mr. Rupert Costo of San Francisco.

LITERATURE CITED

AGINSKY, B. W., and E. G. 1955. Selected papers of B. W. and E. G. Aginsky. Printing Unlimited, New York.

ALLEN, THADDEUS W. 1968. Aboriginal agriculture in the Coachella Valley of the Colorado Desert. 13 pp. [Mimeo.]

ANTHONY, FRANCES. 1900. To Palm Canyon. Land of Sunshine 13(4):237-240.

ASCHMANN, HOMER. 1959a. The central desert of Baja California: demography and ecology. Ibero-Americana 42:282 pp.

———— 1959b. The evolution of a wild landscape and its persistence in Southern California. Ann. Assoc. Amer. Geog. 49(3):II, 34-57.

———— 1966. The head of the Colorado delta. Pp. 231-264. In: Eyre, S. R., and G. R. J. Jones. Geography as human ecology. St. Martin's Press, New York.

ATWATER, W. O., and A. P. BRYANT. 1902. The chemical composition of American food materials. U. S. Dept. Agri. Bul. 28.

BALLS, EDWARD K. 1965. Early uses of California plants. California Natural History Guide No. 10. University of California Press, Berkeley and Los Angeles.

BARROWS, DAVID PRESCOTT. n.d. Unpublished field notes on the Cahuilla Indians on file at the Bancroft Library, University of California, Berkeley.

———— 1900. The ethno-botany of the Coahuilla Indians of southern California. University of Chicago Press, Chicago.

BAUMHOFF, MARTIN A. 1963. Ecological determinants of aboriginal California populations. Univ. Calif. Pubs. Amer. Archaeol. Ethnol. 49:155-236.

BEAN, LOWELL JOHN. 1961. The ethnobotanical report sheet. Archaeol. Surv. Ann. Rpt., Univ. Calif., Los Angeles, 1960-61:233-236.

———— 1964. Cultural change in Cahuilla religious and political leadership patterns. Pp. 1-10. In: Beals, Ralph L. (ed.). Cultural change and stability essays in memory of Olive Ruth Barker and George C. Barker. Department of Anthropology, University of California, Los Angeles.

———— 1970. Cahuilla Indian cultural ecology. University Microfilms, Ann Arbor, Michigan. [Doctoral dissertation, University of California, Los Angeles.]

———— 1972. Mukat's people: the Cahuilla Indians of southern California. University of California Press, Berkeley and Los Angeles.

BEAN, LOWELL JOHN, and HARRY W. LAWTON. 1967. A bibliography of the Cahuilla Indians of California. Malki Museum Press, Morongo Indian Reservation, Banning.

BEAN, LOWELL JOHN, and WILLIAM MARVIN MASON. 1962. Diaries & accounts of the Romero expeditions in Arizona and California, 1823-1826. Palm Springs Desert Museum, Palm Springs.

BEAN, LOWELL JOHN, and KATHERINE SIVA SAUBEL. 1961. Cahuilla ethno-botanical notes: the aboriginal uses of oak. Archaeol. Surv. Ann. Rpt., Univ. Calif., Los Angeles, 1960-61:237-245.

———— 1963. Cahuilla ethnobotanical notes: aboriginal uses of mesquite and screwbean. Archaeol. Surv. Ann. Rpt., Univ. Calif., Los Angeles, 1962-63:51-78.

BEARDSLEY, R. K., et al. 1956. Functional and evolutionary implications of community patterning. In: Wauchope, R. (ed.) Seminars in archaelogy: 1955. Memoir No. 11, Society for American Archaeology, Salt Lake City.

BEATTIE, GEORGE W., and HELEN PRUITT. 1951. Heritage of the valley. San Bernardino's first century. Biobooks, Oakland.

BELL, WILLIS H., and EDWARD F. CASTETTER. 1937. The utilization of mesquite and screwbean by the aborigines in the American Southwest. Ethnobiol. Studies in the Amer. Southwest V. Bul., Biol. Ser., Univ. New Mexico 4(5):3-63.

BENNETT, MELBA. n.d. Ethnographic field notes.

BENSON, LYMAN, and ROBERT A. DARROW. 1944. A manual of southwestern desert trees and shrubs. Univ. Ariz. Bul. 15(2):411 pp.

BLAKE, WILLIAM P. 1953. First wagons across the badlands. Calico Print 9(2):25-27.

BOLTON, HERBERT E. 1907. Spanish explorers in the southern United States, 1528-1543. Charles Scribner's Sons, New York.

———— 1916. Spanish exploration in the Southwest, 1542-1700. Charles Scribner's Sons, New York.

———— 1913. Anza's California expeditions. University of California Press, Berkeley and Los Angeles. 5 vol.

BOSCANA, GERONIMO. 1933. Chinigchinich: A revised and annotated version of Father Geronimo Boscana's historical account of beliefs, usages, customs, and extravagancies of the Indians of this Mission of San Juan Capistrano called the Acagchemem tribe. Translated and annotated by J. P. Harrington. Fine Arts Press, Santa Ana.

BOTKIN, C. W. 1940. Pinon nuts as a food crop. New Mexico Agri. Exper. Sta. Bul. 899.

BOWERS, STEPHEN. 1888. The Conchilla Valley and the Cahuilla Indians. Santa Buenaventura.

BRIGHT, WILLIAM. 1967. The Cahuilla language. Pp. xxi-xxix. In: Barrows, David Prescott. The ethno-botany of the Coahuilla Indians of southern California. Malki Museum Press, Morongo Indian Reservation, Banning.

BROWN, JOHN S. 1923. The Salton Sea region, California: A geographic, geologic, and hydrologic reconnaissance with a guide to desert watering places. U. S. Geol. Surv. Water Supply Pap. 497:292 pp.

CAMPA, A. L. 1932. Pinon as an economic and social factor. New Mexico Business Rev. 1:144-147.

CASTANEDA, CARLOS. 1968. The teachings of Don Juan: A Yaqui way of knowledge. University of California Press, Berkeley and Los Angeles.

———— 1971. A separate reality: Further conversations with Don Juan. Simon and Schuster, New York.

CASTETTER, EDWARD F., and WILLIS H. BELL. 1938. The early utilization and distribution of agave in the American Southwest. Ethnobiol. Studies in the Amer. Southwest VI. Bul., Biol. Ser., Univ. New Mexico 5(4):3-92.

———— 1941. The utilization of yucca, sotol, and beargrass by the aborigines of the American Southwest. Univ. New Mexico Bul. 372:74 pp.

———— 1942. Pima and Papago Indian agriculture. Inter-American Studies I. Albuquerque, New Mexico.

———— 1951. Yuman Indian agriculture: Primitive subsistence on the lower Colorado and Gila Rivers. University of New Mexico Press, Albuquerque.

CAUGHEY, JOHN WALTON. (ed.). 1952. The Indians of southern California in 1852: The B. D. Wilson report and a selection of contemporary comment. Huntington Library, San Marino.

CHASE, J. SMEATON. 1919. California desert trails. Houghton Mifflin Company, New York.

CURTIS, EDWARD S. 1926. The North American Indian. Norwood, Massachusetts. Vol. 15.

DOWNS, JAMES F. 1966. The two worlds of Washoe: An Indian tribe of California and Nevada. Holt, Rinehart, and Winston, Inc., New York.

———— 1966. The significance of environmental manipulation in the Great Basin cultural development. Pp. 29-56. In: d'Azevedo, Warren L., et al. The current status of anthropological research in the Great Basin: 1964. Desert Research Institute, Technical Report Series, No. 1, Reno, Nevada.

DRUCKER, PHILIP. 1937. Culture element distributions. V. Southern California. Univ. Calif. Pubs. Anthrop. Rec. 1:1-52.

ENGLEHARDT, ZEPHYRIN. 1920. San Diego Mission. James H. Barry Co., San Francisco.

———— 1927. San Gabriel Mission and the beginnings of Los Angeles. Mission San Gabriel, San Gabriel.

EZELL, PAUL. 1961. The hispanic acculturation of the Gila River Pimas. Mem. Amer. Anthrop. Assn. No. 90, Menasha, Wisconsin.

FORBES, JACK. 1963. Indian horticulture west and northwest of the Colorado River. Jour. West 2:1-14.

FORDE, C. DARYL. 1931. Ethnography of the Yuma Indians. Univ. Calif. Pub. Amer. Archaeol. Ethnol. 28:83-278.

FOSTER, L. 1916. The feeding value of the mesquite bean. New Mexican Farm Courier, Vol. 4, No. 9.

GARCES, FRANCISCO. 1963. A record of travels in Arizona and California, 1775-1776. A new translation by John Galvin. Howell Books, San Francisco.

GARCIA, F. 1917. Mesquite beans for pig feeding. New Mexico Agr. Exper. Sta. Ann. Rept. 28.

GIFFORD, EDWARD W. 1932. The southeastern Yavapai. Univ. Calif. Pubs. Amer. Archaeol. Ethnol. 29:177-252.

GLENDENNING, ROBERT M. 1949. The Coachella Valley, California: Some aspects of agriculture in a desert. Pap. Mich. Acad. Sci., Arts and Letters 35:173-188.

GOLDSCHMIDT, WALTER. 1959. Man's way. A preface to understanding of human society. Henry Holt and Company, Inc. 253 pp.

HALL, H. M. 1902. A botanical survey of the San Jacinto Mountains. Univ. Calif. Pubs. Bot. 1:139.

HALL, H. M. and J. GRINNELL. 1919. Life zone indicators in California. Proc. Calif. Acad. Sci., Ser. 4. 9(2):37-67.

HARDIN, JAMES W., and JAY M. ARENA. 1969. Human poisoning from native and cultivated plants. Duke University Press, Durham.

HARRINGTON, J. P. 1934. A new original version of Boscana's historical account of the San Juan Capistrano Indians of southern California. Smithsonian Miscellaneous Collections, Vol. 92, No. 4.

———— n.d. Unpublished field notes on file at the Smithsonian Institute, Washington, D.C.

HAVARD, VALERY. 1884. The mesquite. Amer. Nat. 18(5):451-459.

———— 1895. The food plants of the North American Indians. Bul. Torrey Bot. Club. 23:33-46.

HAYES, BENJAMIN. 1929. Pioneer notes from the diaries of Judge Benjamin Hayes. Privately printed, Los Angeles.

HEIZER, R. F. and M. A. WHIPPLE. 1957. The California Indians. University of California Press, Berkeley and Los Angeles.

HEIZER, R. F., and HENRY RAPPAPORT. 1962. Identification of *Distichlis* salt. Masterkey 36(4):146-148.

HENDERSON, RANDALL. 1951. Wild palms of the California desert. Desert Magazine, Palm Desert.

HICKS, FREDERICK. 1961. Ecological aspects of aboriginal culture in the western Yuman area. Ph.D. dissertation, University of California, Los Angeles.

HOOPER, LUCILLE. 1920. The Cahuilla Indians. Univ. Calif. Pubs. Amer. Archaeol. Ethnol. 16:316-380.

JAEGER, EDMUND C. 1957. The North American Deserts. Stanford University Press, Palo Alto.

———— 1958. Desert wild flowers. Stanford University Press, Palo Alto.

JAMES, GEORGE WHARTON. 1906. Wonders of the Colorado Desert. Little, Brown & Co., Boston. 2 vol.

KROCHMAL, A., S. PAUR, and P. DUISBERG. 1954. Useful native plants in the American southwestern deserts. Economic Bot. 8(1):3-20.

KROEBER, A. L. 1908. Ethnography of the Cahuilla Indians. Univ. Calif. Pubs. Amer. Archaeol. Ethnol. 6:29-68.

———— 1925. Handbook of the Indians of California. Bureau of American Ethnology, Washington, D.C. Bul. 78: 995 pp.

LAWTON, HARRY W. 1967. The dying god of the Cahuilla: ethnohistoric evidence of a Colorado River-derived agricultural complex in southern California. Graduate seminar paper in English 275A. The Oral Epic., University of California, Riverside.

LAWTON, HARRY W., and LOWELL JOHN BEAN. 1968. A preliminary reconstruction of aboriginal agricultural technology among the Cahuilla. The Indian Historian 1(5):18-24, 29.

LeCONTE, JOHN L. 1855. Account of some volcanic springs in the desert of the Colorado in southern California. Amer. Jour. Sci. Arts 19:1-6.

LEWIN, LOUIS. 1964. Phantastica, narcotics, and stimulating drugs. E. P. Dutton & Company, New York.

MARTINEZ, MAXIMO. 1944. Las plantas medicinales de Mexico. Ediciones Botas, Mexico.

MEIGHAN, CLEMENT. 1959. California cultures and the concept of an archaic stage. Amer. Antiq. 24(3):289-305.

MERRILL, RUTH EARL. 1923. Plants used in basketry by the California Indians. Univ. Calif. Pubs. Amer. Archaeol. Ethnol. 20:215-242.

MORTON, JULIA. 1963. Principal wild food plants of the United States, excluding Alaska and Hawaii. Econ. Bot. 17(4):319-330.

MUNZ, PHILIP A., and DAVID P. KECK. 1968. A California flora. University of California Press, Berkeley and Los Angeles.

PALMER, EDWARD. 1871. Food products of the North American Indians. U. S. Dept. Agr. Rept. *1870*:404-428.

———— 1878. Plants used by the Indians of the United States. Amer. Nat. *12*:593-606; 646-655.

PATENCIO, FRANCISCO. 1943. Stories and legends of the Palm Springs Indians. Times-Mirror Press, Los Angeles.

PRICE, JOHN A. 1968. Abstracts: Southwestern Anthropology Association and Society for California Archaeology, 1969. San Diego. [Mimeo.]

REED, LESTER. 1963. Old time cattlemen and other pioneers of the Anza-Borrego area. Privately printed, Palm Desert.

ROGERS, MALCOLM. 1938. The aborigines of the desert. Pp. 116-129. *In*: Jaeger, Edmund C. The California deserts. Stanford University Press, Palo Alto.

ROMERO, JOHN BRUNO. 1954. The botanical lore of the California Indians with sidelights on historical incidents in California. Vantage Press, Inc., New York.

RUIBAL, RODOLFO. 1959. The ecology of a brackish water population of *Rana pipiens*. Copeia 1959(4):315-322.

SAUNDERS, CHARLES FRANCIS. 1913. Under the sky in California. McBride, Nast & Company, New York.

———— 1914. With the flowers and trees in California. Robert M. McBride & Company, New York.

SOLLMAN, TORALD. 1957. A manual of pharmacology and its applications to therapeutics and toxicology. W. B. Saunders Company, New York.

SPARKMAN, PHILIP STEDMAN. 1908. The culture of the Luiseño Indians. Univ. Calif. Pubs. Amer. Archaeol. Ethnol. 8:187-234.

SPENCER, EDWIN ROLLIN. 1949. Just weeds. Charles Scribner's Sons, New York.

SPENCER, WILLIAM S. 1956. Handbook of biological data. W. B. Saunders Company, New York.

SPIER, LESLIE. 1933. Yuman tribes of the Colorado River. University of Chicago Press, Chicago.

STEWARD, JULIAN H. 1929. Irrigation without agriculture. Pap. Mich. Acad. Sci., Arts and Letters 22:149-156.

———— 1955. The theory of culture change. University of Illinois Press, Urbana.

———— 1938. Basin-plateau aboriginal socio-political groups. Smithsonian Inst. Bur. Amer. Ethnol. Bul. *120*:346 pp.

STICKNEY, FENNER S., DWIGHT F. BARNES, and PEREZ SIMMONS. 1950. Date palm insects in the United States. U. S. Dept. Agr. Circ. 846.

STRONG, WILLIAM DUNCAN. 1927. An analysis of southwestern society. Amer. Anthrop., n.s., 29:1-6.

———— 1929. Aboriginal society in southern California. Univ. Calif. Pubs. Amer. Archaeol. Ethnol. 26:329 pp.

SUDWORTH, GEORGE B. 1908. Forest trees of the Pacific slope. U. S. Dept. Agri. Forest Service, Washington, D.C.

TREGANZA, ADAN E. 1947. Possibilities of an aboriginal practice of agriculture among the Southern Diegueño. Amer. Antiq. 12:169-173.

UNDERHILL, RUTH. 1936. The autobiography of a Papago woman. Mem. Amer. Anthrop. Assoc. 46.

VEATCH, JOHN A. 1858. Notes of a visit to the "mud volcanoes" in the Colorado Desert in the month of July, 1857. Amer. Jour. Sci. Arts 27: 288-295.

WHIPPLE, A. W. 1855. Report of explorations for a railway route near the thirty-fifth parallel of north latitude from the Mississippi River to the Pacific Ocean, 1853-1854. Vol. III. U. S. Government Printing Office, Washington, D.C.

WHITE, RAYMOND C. 1963. Luiseño social organization. Univ. Calif. Pubs. Amer. Archaeol. Ethnol. 48(2):194 pp.

WILKE, P. J., and DOUGLAS FAIN. 1972. An archaeological cucurbit from Coachella Valley, California. Archaeological Research Unit, Department of Anthropology, University of California, Riverside. 3 pp. [Mimeo.]

WILLEY, GORDON R., and PHILIP PHILLIPS. 1958. Method and theory in American archaeology. University of Chicago Press, Chicago.

WILLIAMSON, LT. R. S. 1856. Report of explorations in California for railroad routes to connect with routes near the 35th and 32d parallels of north latitude. Vol. V. Beverly Tucker, Printer, Washington, D.C.

WOLF, CARL B. 1945. California wild tree crops. Rancho Santa Ana Botanic Garden of the Native Plants of California.

ZIGMOND, MAURICE LOUIS. 1941. Ethnobotanical studies among California and Great Basin Soshoneans. Ph.D. dissertation on file at Yale University.

Index of Scientific Names*

*Page numbers specifically devoted to a plant are given first, and these are followed by more general references.

220

Index of Common Names